Learning and Analytics in Intelligent Systems

Volume 25

The main aim of the series is to make available a publication of books in hard copy form and soft copy form on all aspects of learning, analytics and advanced intelligent systems and related technologies. The mentioned disciplines are strongly related and complement one another significantly. Thus, the series encourages cross-fertilization highlighting research and knowledge of common interest. The series allows a unified/integrated approach to themes and topics in these scientific disciplines which will result in significant cross-fertilization and research dissemination. To maximize dissemination of research results and knowledge in these disciplines, the series publishes edited books, monographs, handbooks, textbooks and conference proceedings.

More information about this series at https://link.springer.com/bookseries/16172

Steven Lawrence Fernandes · Tarun K. Sharma
Editors

Artificial Intelligence in Industrial Applications

Approaches to Solve the Intrinsic Industrial Optimization Problems

Springer

Editors
Steven Lawrence Fernandes
Department of Computer Science
Creighton University
Omaha, NE, USA

Tarun K. Sharma
Shobhit University
Saharanpur, India

ISSN 2662-3447 ISSN 2662-3455 (electronic)
Learning and Analytics in Intelligent Systems
ISBN 978-3-030-85382-2 ISBN 978-3-030-85383-9 (eBook)
https://doi.org/10.1007/978-3-030-85383-9

This Springer imprint is published by the registered company Springer Nature Switzerland AG
The registered company address is: Gewerbestrasse 11, 6330 Cham, Switzerland

Preface

This book highlights the analytics and optimization issues in the industry, proposes new approaches, and presents applications of innovative approaches in real facilities. In the past few decades, there has been an exponential rise in the application of artificial intelligence for solving complex and intricate problems arising in the industrial domain. The versatility of these techniques has made them a favorite among scientists and researchers working in diverse areas.

The present book includes only 12 chapters having the applications of intelligent algorithms in the Industry.

In chapter "An Efficient Deep Learning Framework for People Detection in Overhead Images", Peker et al. propose an efficient deep learning framework for detecting people from overhead images. The proposed method uses a multi-scale Yolov4-Tiny algorithm for person detection. The multi-scale feature of the algorithm enables it to successfully detect people who appear small in an image. The proposed method not only shows promising results, but can also work in real-time applications due to low computational cost.

In chapter "Machine Learning Techniques for Grocery Sales Forecasting by Analyzing Historical Data", Yeasmin et al. proposed a forecasting model using machine learning algorithms, to get accurate forecasts Grocery Sales Forecasting by Analyzing Historical Data.

In the chapter "Medication Revelation Utilizing Neural Network", Kushwah et al. present a review of the applications that produce promising results and methods will be reviewed. The use of virtual screening and encompassing online information has also been highlighted in developing lead synthesis pathways.

In the chapter "Application of AI in SCM or Supply Chain 4.0", Singh et al. provide the speculative allusion of application of AI in SCM, reimbursement, and the challenges of AI in the SCM industry. In chapter "Upskilling and Curating the Potentials of IoT Enabled Smart Cities: Use Cases and Implementation Strategies", Sivarethinamohan and Sujatha examine significant aspects of an IoT infrastructure for smart cities, outlining the innovations implemented in the cities of India as use cases and Implementation Strategies. Exceptional attention is devoted to the potential applications of smart cities. Datta and Mitra, in chapter "Identifying Impact of Sanitation and Environmental Safety in Hospitality Sector to Identify the Spread of COVID-19 Using Polynomial Function Simulation", detailed and solved a challenging problem due to identifying the impact of sanitation and environmental safety in the hospitality sector to identify the spread of COVID-19 using Polynomial Function Simulation. The chapter "The Internet of Things and Advanced Applications in Healthcare", by Gandhi and Singh, addressed IoT technology, IoT architecture, and IoT advanced applications in healthcare. They have introduced IoT application in hospital beds that will provide complete patient care and decrease the requirement of the hospital caring staff, and all activities of the patient can be transmitted to the doctor's room or chamber to update them about the patient's current status/ vital signs in ICU. The chapter "A Comprehensive Review of Recent Automatic Speech Summarization and Keyword Identification Techniques", by Kumar et al., aims to outline and explain some of the popular approaches in speech recognition systems at various stages and highlights selected systems' unique and innovative characteristics. In the next chapter "Anomaly Detection in Industrial IoT Applications Using Deep Learning Approach", Bulla and Birje propose an anomaly detection in Industrial IoT Applications using a Deep Learning approach. The proposed anomaly detection scheme uses a multi-step prediction technique and applies an anomaly detection algorithm to detect anomalies. The proposed model increases accuracy over existing approaches, according to the experimental evaluation. In the chapter "Nurturing the Rudiments and Use Cases of Ongoing Natural Language Generation for a Future Profitable Business More Profitable", Sivarethinamohan and Sujatha presented a study of how NLG enables machines and humans to communicate seamlessly, simulating human-to-human conversations, and using NLG how organizations are building new customer experiences, monetizing information assets, introducing new offerings, and streamlining operational costs. Therefore, the coverage of this chapter will answer to the industrialists and new start-ups. What can NLG do for business? And what are the future applications of NLG?. In the chapter "A Study on Deep Learning Models for Medical Image Segmentation", Bhatt et al. highlight the most popular algorithms based on DL architecture with respect to the segmentation of medical images. Bhardwaj in the chapter "Deep Learning-based Cyber Security Solutions for Smart-City: Application and Review" discuss about the Convolutional Neural Networks (CNNs), Fully Connected Convolutional Networks (FCNs), Recurrent. Neural Network (RNN), Deep Belief Network (DBN), Boltzmann Machine, and Auto-encoders.

The book is edited to serve a broad readership, including computer scientists, medical professionals, and mathematicians interested in studying tools and techniques for computational intelligence and applications in healthcare. It will also be helpful for researchers, graduate, and undergraduate students with an interest in the fields of Artificial Intelligence and Industrial problems.

This book will be a useful resource for researchers and academicians as well as professionals interested in the highly interdisciplinary field of Artificial Intelligence.

Omaha, USA — Steven Lawrence Fernandes
Saharanpur, India — Tarun K. Sharma

Contents

Chapter 1
An Efficient Deep Learning Framework for People Detection in Overhead Images

Musa Peker, Bilge İnci, Elnura Musaoğlu, Hüseyin Çobanoğlu, Nadir Kocakır, and Önder Karademir

Abstract People detection is an important issue in video surveillance applications. The factors such as severe occlusions and scene perspective distortions in real application scenarios complicate this task. This article proposes an efficient deep learning framework for detecting people from overhead images. The proposed method uses a multi-scale Yolov4-Tiny algorithm for person detection. The multi-scale feature of the algorithm enables it to successfully detect people who appear small in an image. A comparative analysis was made with non-maximum suppression (NMS) algorithms, which significantly affect the detection performance. In this context, it is aimed to determine the most suitable NMS algorithm for the multi-scale Yolov4-Tiny algorithm. The performance of this method has been evaluated by various experiments. The experiments were carried out on two datasets consisting of many difficulties such as light, distance, occlusion, and wear a hat. As a result of the experiments, 94.85% mean average precision (mAP) value was obtained with the proposed method. The proposed method not only shows promising results, but can also work in real-time applications due to low computational cost.

Keywords Multi-scale Yolov4-Tiny algorithm · Top-view people detection · NMS · Object detection

1.1 Introduction

People detection is one of the important tasks in smart video surveillance systems. It has business value and wide application in many places. For example, it is widely used in many fields such as video surveillance, people counting systems, access control or human behavior analysis. There are many studies in the literature that perform people's detection and tracking using hardware such as cameras and sensors [1–5].

M. Peker (✉) · B. İnci · E. Musaoğlu · H. Çobanoğlu · N. Kocakır · Ö. Karademir
Özdilek Ev Tekstil San. Ve Tic. AŞ, Ozveri R&D Center, Bursa, Turkey
e-mail: ozveri@ozdilek.com.tr

S. L. Fernandes and T. K. Sharma (eds.), *Artificial Intelligence in Industrial Applications*, Learning and Analytics in Intelligent Systems 25,
https://doi.org/10.1007/978-3-030-85383-9_1

Although there are many studies in the literature on person detection, it is a challenging task due to severe blockages, scene distortions, and diverse crowd distributions. Many techniques have been shared in the literature by researchers trying to overcome these difficulties [6, 7]. Most of the techniques developed used front or side view cameras. However, there is a large amount of occlusion problems in these cameras. To solve this problem, some researchers have used multiple cameras to accurately detect humans [8]. But this requires high costs.

When using top view cameras, the occlusion problem can be reduced significantly. Thanks to this feature, these types of cameras are frequently used in industry and academic studies for person detection. In addition, privacy issues are reduced by using these cameras. Because the camera only captures the person's body from the top view instead of the facial images [9]. But some situations such as people walking close to each other (piggybacking, tailgating) in overhead images are still a challenging task.

In the light of the above discussion, a deep learning-based solution has been developed for detecting people from overhead images in this study. The proposed model uses a modified Yolov4-Tiny algorithm. This model has been tested on images obtained at different heights and lighting conditions and its effectiveness has been investigated.

1.1.1 Literature Survey

In the literature, some studies on detecting people from top-view images are as follows. Ahmad et al. [10] developed a method for counting people from overhead images. In the proposed method, a background subtraction based method was used for person detection. Del Pizzo et al. [11] developed a method based on foreground detection to people detection from top-view images. The proposed method is a quick background subtraction color-based algorithm. Mukherjee et al. [12] proposed a new framework for counting passengers in a railway station. The Hough circle-based algorithm has been used to people detection, since the view of the body of the person is circular when viewed from above. In addition, the proposed method has been compared with standard histogram of oriented gradients (HOG) features. Sun et al. [13] proposed a people counting system from overhead images obtained from an RGB-D camera. In the proposed system, a 3D human model is used to people detection. Carletti et al. [14] developed a method for people counting using depth map images. They used a background extraction method for person detection from above. Kristoffersen et al. [15] used thermal camera images for person counting. A Shi–Tomasi corner detector with the scale-invariant feature transform (SIFT) descriptors was used for person detection. In [16] and [17], a feature-based algorithm was used with the support vector machine (SVM) classifier to detect people from top-view images. Ahmed et al. [18] used an algorithm called the rotated histogram of oriented gradient (RHOG), which uses geometric transformations to standardize the different orientations and sizes of bounding boxes. Successful results were obtained in the study in which SVM algorithm was used as a classifier.

From the above discussion, it is seen that top view person detection studies are generally carried out using background subtraction and traditional methods. In the literature, the number of studies on detecting people with deep learning techniques on overhead images is quite low. Ahmad et al. [19] used the YOLOv3 algorithm to detect people from overhead images. The developed model was trained on the frontal view data set. And this model has been tested on the overhead view person data set. With the proposed model, 95% True Positive Rate (TPR) value was obtained. Ahmad et al. [20] proposed a single shot multi-box detector (SSD) based method to detect people from top-view images. With the proposed method, average 95% TPR and 94.42% TPR values were obtained for indoor and outdoor environments, respectively. Ahmed et al. [21] used Faster-RCNN and Mask-RCNN algorithms to detect people from overhead images. These models were trained on the frontal view data set, and tested on the overhead view person data set. With the Faster-RCNN and the Mask-RCNN model, a true positive rate (TPR) of 94 and 93% was achieved, respectively.

1.1.2 Contribution Summary

As seen in the literature study, the use of deep learning-based methods for people detection from overhead images is a new research area. In this study, a deep learning-based model is proposed to people detection from overhead images. The main contributions of the proposed method are listed below.

- A modified version of Yolov4-Tiny algorithm is proposed for person detection from overhead images. The proposed method is named multi-scale Yolov4-Tiny.
- When deep learning-based studies for people detection from overhead images in the literature are examined, it is seen that the developed models are trained on frontal view data sets and tested on top-view datasets. In this study, both training and testing processes were carried out over top-view datasets. This ensures that the created model is specific to top-view images. And this is an important factor in increasing the success rate.
- In this study, comparative performance tests were conducted for NMS algorithms used in post-processing, which significantly affect the success performance of one stage detectors. It is aimed to determine the most suitable NMS algorithm for multi-scale Yolov4-Tiny.
- As a result of the experiments, it has been seen that the proposed method is an effective solution in terms of a trade-off based on inference time and accuracy.

1.2 Dataset

In this study, training and testing processes were carried out on two datasets. These datasets are introduced below.

Dataset 1- Sunmi AI Lab In Office Dataset [22]*:* This dataset is created using a video from a single ceiling-mounted camera in an office environment. The video simulates six different behaviors typical of customers in a supermarket. These behavioral patterns explained in Table 1.1 consist of factors that make it difficult to detecting people from overhead. A few images of this dataset are seen in Fig. 1.1.

Dataset 2- Top view multi person dataset [23]*:* This dataset consists of a total of 4000 images obtained from 5 cameras with a frequency of 24 frames / second and a size of 1280 × 960. In this dataset, people seem smaller because the camera is higher than the ground. A few images of this dataset are seen in Fig. 1.2.

Table 1.1 Overview of the dataset obtained in an office environment with six different behaviors

Behavior	Explanation	Number of images
Wearing hats	Participants wear hats	165
Covering heads	Participants use their hands to cover their heads	817
Squatting	Participants squats as if picking up products from lower shelves	167
Putting jackets on/off	Participants put on and take off their jackets	166
Lights off	Lights are turned off while participants continue to move	145
Close proximity	Participants walk close to each other so their head overlapping	167

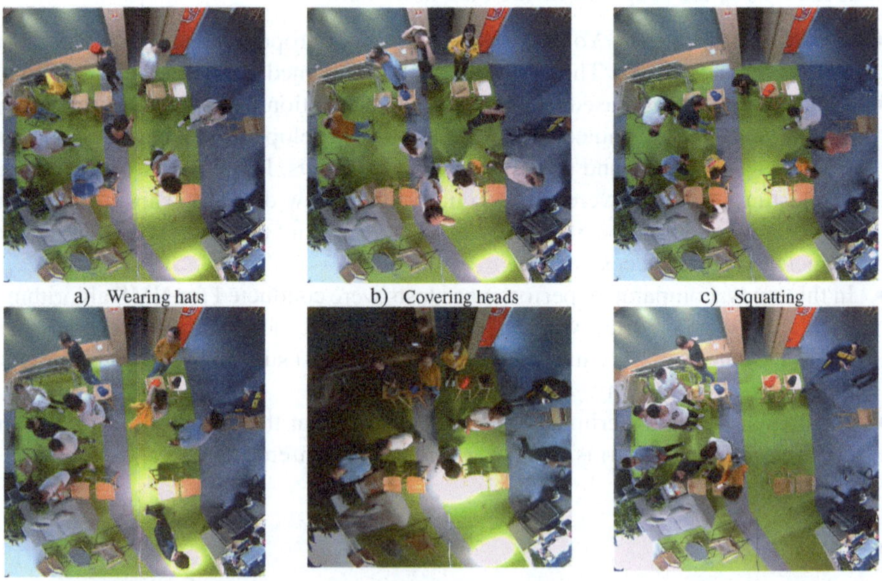

a) Wearing hats b) Covering heads c) Squatting

d) Putting jackets on/off e) Lights off f) Close proximity

Fig. 1.1 Sample images from Dataset 1 for each type of behavior [22]

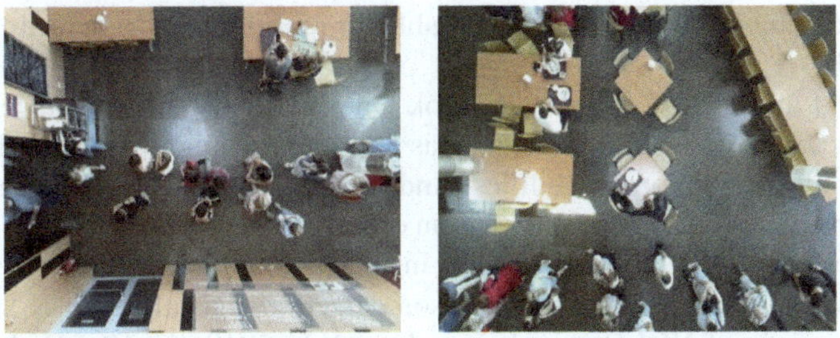

Fig. 1.2 Sample images from Dataset 2 [23]

1.3 Methods

1.3.1 Object Detection Algorithms

Modern deep learning based detection algorithms consist of 3 stages as Backbone, Neck and Head. This is schematically presented in Fig. 1.3. The backbone is expressed as feature extraction architecture [24] and is one of the main factors that increase the accuracy of the system. Some feature extraction architectures are: VGG16 [25], ResNet-50 [26], CSPResNeXt-50 [27], and CSPDarknet53 [27]. The neck section provides quality visual features for the head section. It comprises of several bottom-up and top-down ways to aggregate and combine the parameters of the network in different layers. In a deep learning architecture, the Head section enables to classify objects (such as people, bicycles, and cars). It also calculates the coordinates of bounding boxes and the size of objects. Object detection methods are divided into two categories: one stage detectors and two stage detectors. Although two-stage detectors give successful results, they have high computational costs [28].

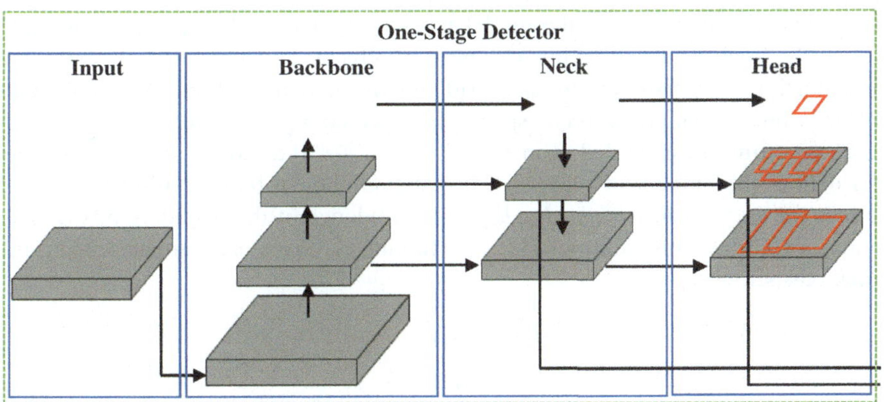

Fig. 1.3 The structure of one stage detectors [32]

One stage detectors, which are more suitable for real-time object detection, are used in this study.

SSD [29] and YOLO (You Only Look Once) [30–32] are one stage detectors. These detectors utilize regression analysis to compute the dimensions of bounding boxes and interpret class probabilities. And these detectors map the image pixels to the grids and control the probability that an object exists in each cell of the grids. This approach offers excellent improvements in speed and efficiency [24]. In this study, YOLO based algorithm is used as a person detection algorithm. Since 2016, many different versions of YOLO have been released. In 2018, the YOLOv3 algorithm, an improved version of YOLOv2, was proposed [31]. Compared to the Darknet-19 architecture used by Yolov2, YOLOv3 uses the Darknet-53 architecture consisting of 53 convolution layers to deepen the network structure and residual blocks have been added to this network. An important feature of YOLOv3 is the multi-scale prediction, which improves the algorithm's ability to predict small objects. The YOLO4 algorithm, an advanced version of YOLOv3, was introduced in 2020 [32]. This algorithm integrates some new technologies such as weighted residual networks, cross-stage partial networks, and cross-mini-batch normalization to improve the speed and accuracy of object detection.

In addition, Yolov3 and Yolov4 were launched with a simplified and optimized structure without much loss of accuracy. This structure is named Tiny. In this study, a Yolov4-Tiny based algorithm is used, which offers significant advantages in terms of both accuracy and speed. This algorithm includes a number of modifications to improve the existing Yolov4-Tiny architecture. In the experimental results section, a comparative performance analysis was made using Yolov3, Yolov4, Yolov3-Tiny and Yolov4-Tiny algorithms.

1.3.2 The Detection Approach of Yolo-Based Algorithms

As shown in Fig. 1.4, after the model takes an image, it divides the image into *SxS* grids. Each grid cell estimates only one object, and image localization and classification are performed in each grid. When the center of an object coincides with a grid cell, that grid cell is responsible for detecting that object.

These estimates are coded as an $SxSx[Bx(4+1+C)]$ tensor for the whole image. *SxS* defines the number of grid cells. B is the number of bounding boxes. $(4 + 1)$ defines the coordinates (b_x, b_y, b_w, b_h) and confidence of the bounding box (p_c). C represents C class probabilities (i.e. c_1). Boundary boxes are estimated using the equations below.

$$\begin{aligned} b_x &= \sigma(t_x) + c_x \\ b_y &= \sigma(t_y) + c_y \\ b_w &= p_w e^{t_w} \\ b_h &= p_h e^{t_h} \end{aligned} \tag{1.1}$$

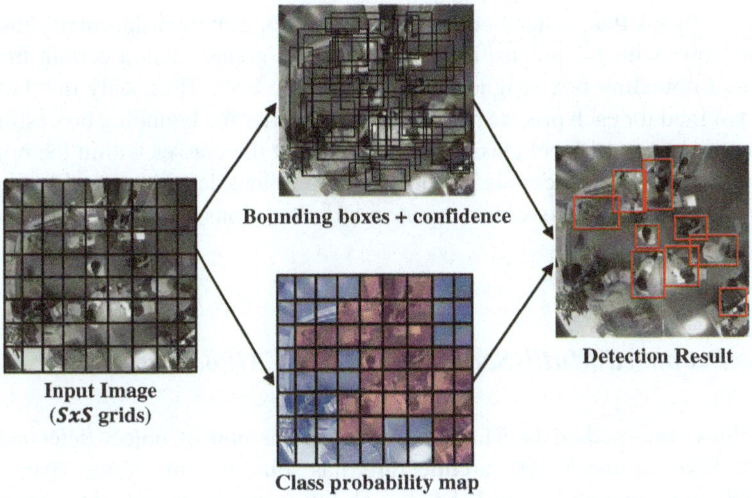

Fig. 1.4 The model splits the image to *SxS* block. And then, bounding boxes containing persons are estimated, and the probability of a class for person is also estimated. With these two information, the person detection process is carried out

Here t_x, t_y, t_w, t_h are the estimated values of the network. σ represents the sigmoid function. p_w and p_h are the dimensions of the specified anchor boxes. c_x and c_y are used to determine the position of the boundary box. Using these values, the position of the boundary box (b_x and b_y) and the dimensions of the boundary box (b_w, b_h) are calculated. It is shown in detail in Fig. 1.5.

In YOLO based models, an objectivity score for each bounding box is predicted by utilizing logistic regression. It should be 1 if a class box contains more precise

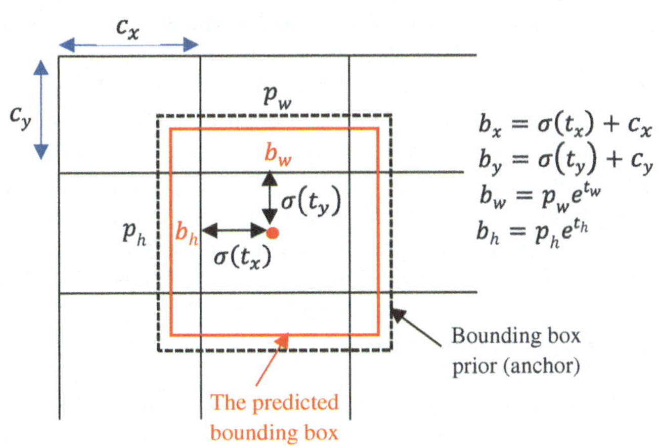

Fig. 1.5 Bounding box prediction with specifications [30]

reference objects than other boundary boxes. Also, although the intersection of a bounding box with the precise reference object is greater than a certain threshold value, that bounding box is ignored if it is not the best. Thus, only one bounding box is assigned for each precise reference object. After the bounding box estimation, each box uses a multi-label classification to predict the classes within the bounding box. Post-processing process is performed for the final detection. At this stage, the NMS method is utilized to remove unnecessary detections and find the most suitable match.

1.3.3 Proposed Multi-Scale Yolov4-Tiny Algorithm

The Yolov4-tiny method is a fast and effective method of object detection. This method, based on the Yolov4 architecture, has some optimizations. Thanks to its affordable computational cost, it has the ability to work on embedded systems or mobile devices.

CSPDarknet53 architecture is used as the backbone structure in the Yolov4 method. In Yolov4-Tiny, CSPDarknet53-Tiny architecture is used as the backbone network. In the CSPDarknet53-tiny network, the residual block module in the residual network [33] is not used. Instead, there is the cross stage partial connections block (CSPBlock) module in the cross-stage partial network.

CSPBlock structure is shown in Fig. 1.6. After the CSPBlock module seperates the feature map into two parts, it combines these parts with the cross stage residual edge. This allows the gradient flow to travel through two different network paths. This increases the correlation difference of gradient information. This feature of the CSPBlock module can positively affect the learning ability of the convolutional network [33].

In the Yolov4 algorithm, the Mish function is utilized as the activation function. To reduce the computation cost, the Leaky ReLU function is used in place of the Mish function in the Yolov4-Tiny algorithm. In the feature fusion part, the Yolov4-tiny method follows a different strategy than the Yolov4 method. Unlike the path aggregation network and spatial pyramid pooling used in Yolov4, it uses the feature pyramid network to increase the speed of object detection. Yolov4-tiny utilizes maps with two different scale attributes, 13×13 and 26×26, to predict detection results. In this work, some improvements were made to increase the success performance of the Yolov4-Tiny algorithm. The changes made are stated below.

Multi-scale Yolov4-Tiny: For the input image in the Yolov4-Tiny algorithm, the size of the feature maps of the two detection layers is 13×13 and 26×26, respectively. As the size of the feature map decreases, it will grow in the area to which each grid cell in the input image will correspond. Conversely, the larger the size of the feature map, the smaller the area that each grid cell in the input image will correspond to. This means that the 13×13 detection layer is appropriate for detecting large targets, while the 26×26 detection layer is appropriate for detecting smaller targets.

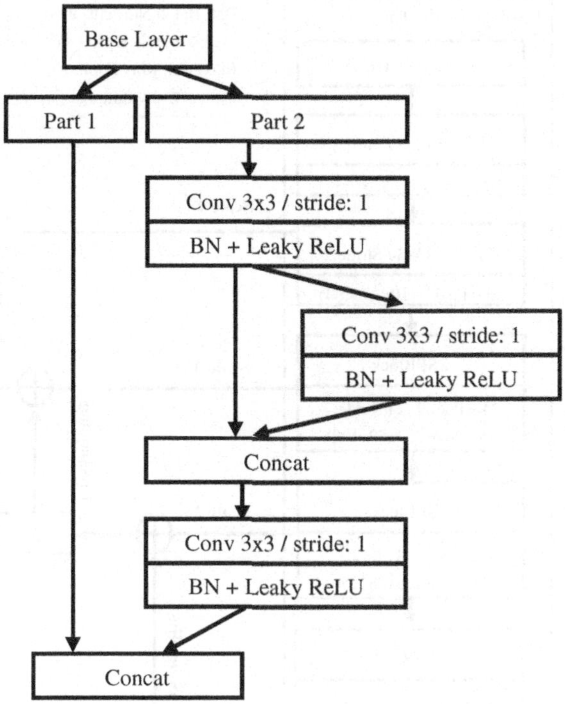

Fig. 1.6 The structure of CSPBlock [33]

In overhead images, people will appear smaller as the camera moves away from the ground. As the height increases, a larger sized detection scale will be needed to further increase the detection performance. For this purpose, two new scales have been added to the existing Yolov4-Tiny architecture in this study. These are 56 × 56 and 104 × 104 respectively. Compared to the original two scales, the four detection scale approach is appropriate for detecting targets with smaller sizes. This proposed multi-scale Yolov4-Tiny architecture is shown in Fig. 1.7. The regions highlighted in red indicate additions to the original Yolov4-Tiny algorithm.

NMS: One stage detectors allow multiple predictions of the same object in a single image. These predictions then become clear with a process called non-maximum suppression (NMS), which allows finding only the highest probability bounding box and label for the object [34].

When the studies using Yolov4-Tiny are examined, it is seen that the basic NMS algorithm is generally used in the post-processing phase.

The predictions of a model in the image detection task are based on the following factors: (1) the coordinates of the bounding box rectangle, (2) a class label of the object, and (3) a confidence score (probability from 0 to 1) reflecting how accurate the model is in this prediction. In the case of boxes overlapping in the NMS method, if the intersection-over union (IoU) is higher than the specified threshold values, these boxes are considered to belong to a single object. Therefore, the process of filtering boxes depends on the choice of this IoU threshold, which affects the performance

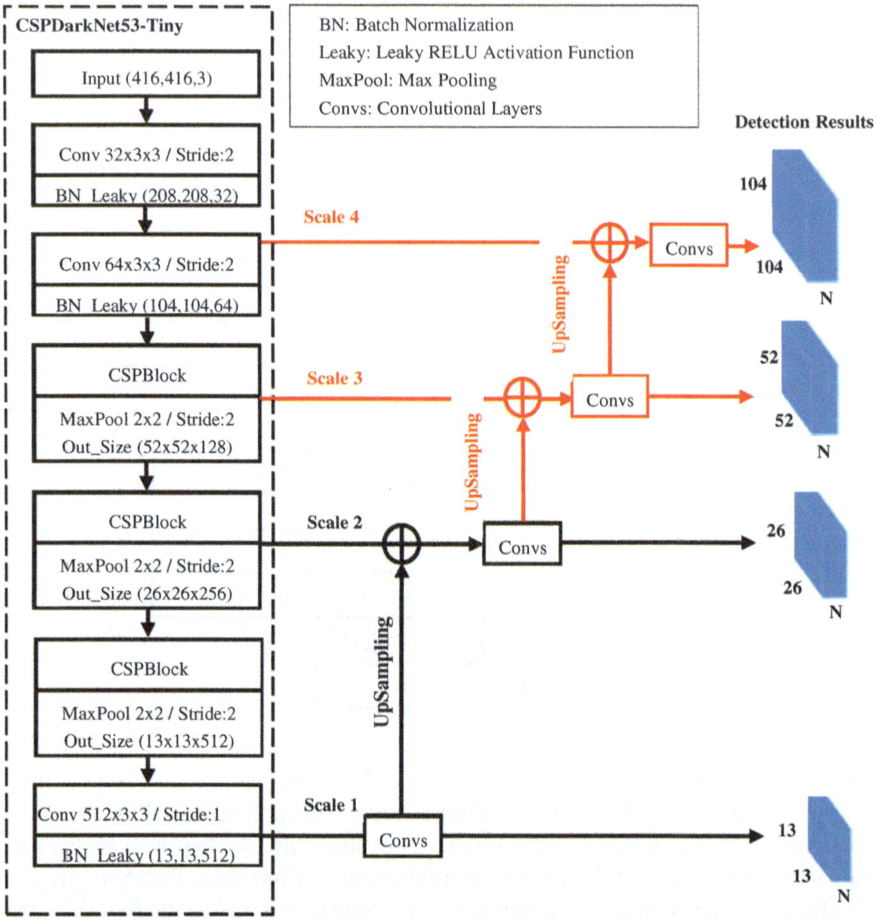

Fig. 1.7 Multi-scale Yolov4-Tiny algorithm

of the model. However, this threshold is difficult to determine. If there are objects next to each other, one of them will be eliminated. In Fig. 1.8, such a situation is simulated. Let's consider the yellow and red colored boxes. Only one box estimate will remain for the 0.5 IoU threshold. These types of errors reduce the precision of the model.

There are many NMS algorithms developed in the literature to eliminate the above-mentioned disadvantage. Soft Non-Maximum Suppression (Soft-NMS) [35] is used to improve the performance of the NMS. It uses the linear penalty function to lower the detection scores of all other neighbors, thus increasing the detection performance. DIoU-NMS [34] takes into account not only the overlapping areas but also the distance between the center points between two borders.

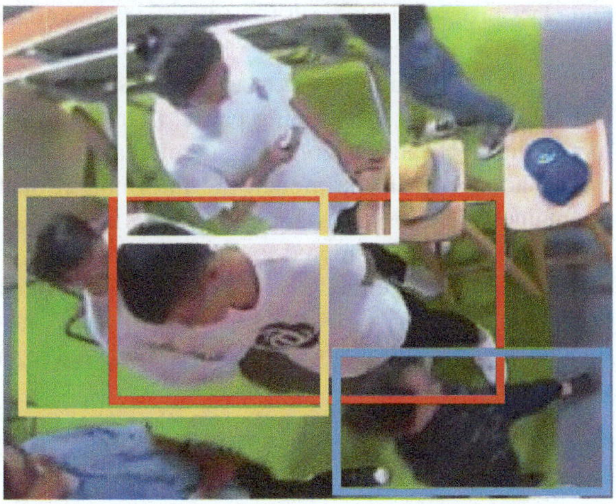

Fig. 1.8 This image is obtained by zooming in the overlapping area in an image taken from Dataset 1. Depending on the IOU threshold value determined, one of the yellow and red rectangles may be eliminated due to the weakness in the NMS algorithm

In this study, it is aimed to determine the most suitable NMS algorithm for the multi-scale Yolov4 Tiny by conducting experiments with these different NMS algorithms.

1.4 Experimental Setup

The detectors are based on Python 3.6, Tensorflow 2.0 and Darknet framework in Windows 10. All experiments were conducted using a computer with the following features: Intel® Core™ i9-9900 K CPU, 32 GB memory, and NVIDIA GeForce RTX 2080 Ti GPU.

The proposed model in this study was compared with four Yolo-based models. These are YOLOv3, YOLOv3-tiny, YOLOv4 and YOLOv4-tiny.

1.4.1 Training Stage

All YOLO based algorithms are trained using stochastic gradient descent (SGD) optimization with Adam optimizer. To achieve better detection performance as the size of the input image increases, YOLO-based algorithms were trained with input image sizes (416 × 416) and (608 × 608). The values of the parameters used for training are as follows: the initial learning rate is 0.001; the iteration is 50.000; the batch and the subdivision sizes are 64 and 24, respectively; the momentum and weight decay are respectively as 0.85 and 0.0005. As data augmentation parameters, saturation value = 1.5, hue = 0.1 and exposure = 1.5 were used. During the training, 0.5 IOU threshold and 0.25 confidence threshold were used to update the regression and class probability loss. 80% of the data in the dataset was used as a training set and the 20% as a test set.

Details of the training and testing processes of the proposed model are presented in Fig. 1.9. During the training phase, training images are presented as data to the model through data resizing and augmentation. In the next step, with the help of the anchor boxes in the model, the information of the estimated bounding boxes is obtained. The loss between the ground truth bounding boxes and the predicted bounding boxes is computed to complete a training period. Many training sessions repeat until the predefined number of iterations is reached. In the detection stage, the estimated bounding boxes using the trained model are obtained. At the last stage, the final detection result is obtained. NMS is utilized to find the best match and remove unnecessary detections.

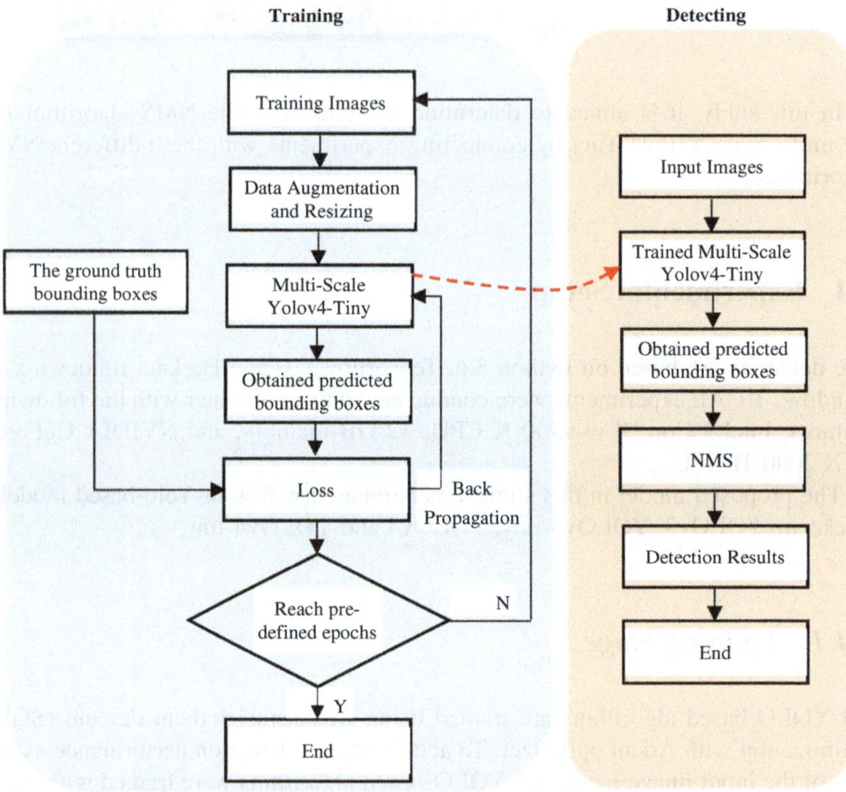

Fig. 1.9 Training and detection process of the multi-scale Yolov4-Tiny model

1.4.2 Model Evaluation Metrics

This article uses some metrics to quantitatively evaluate the success of the proposed model. Information about the performance evaluation methods used in this study is given below.

Precision (P) and Recall (R) rate: Precision defines to the proportion of true positive (TP) results in the predicted positive sample, and recall defines to the proportion of TP results in the truly positive sample. Precision evaluates the ability of a model to accurately and specifically describe an object. Recall evaluates a model's ability to find all ground truths. These two parameters are calculated as shown in Eqs. 1.2 and 1.3.

$$P = \frac{TP}{FP + TP} \tag{1.2}$$

$$R = \frac{TP}{FN + TP} \tag{1.3}$$

where TP, TN, FN and FP define the number of true positives, true negatives, false negatives and false positives, respectively.

F1 score and mean average precision (mAP): The mAP is a performance evaluation parameter that takes into account the recall rate and precision. The mAP value, which expresses the mean value of the mean precision (AP) of all classes, is equal to AP due to a single object (person). As seen in Eq. 1.4. AP is expressed by the area under the $(precisexrecallrate)$ curve.

$$AP = \int_{0}^{1} P(R)dR \tag{1.4}$$

One of the criteria used to evaluate the performance of the models is the F1 score. The F1 score is calculated as seen in Eq. 1.5.

$$F1 = \frac{2P.R}{P + R} \tag{1.5}$$

FPS and inference time: FPS refers to the number of frames transmitted per second in the image processing area. Inference time refers to the time required to process an image. In this study, both are utilized to predict model processing speed.

1.5 Experimental Results and Discussion

The visual results of the test results obtained with the proposed method are presented in Figs. 1.10 and 1.11. Undetected persons or false detections are indicated by a red

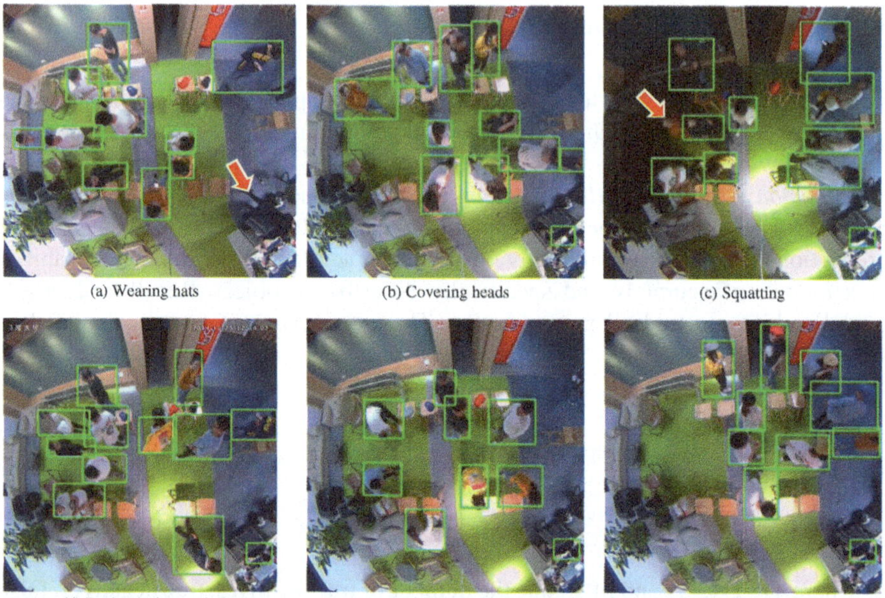

(a) Wearing hats (b) Covering heads (c) Squatting

(d) Putting jackets on/off (e) Lights off (f) Close proximity

Fig. 1.10 Visual results of the proposed method on Dataset 1

arrow. When the figures are examined, it can be said that the proposed method shows a good performance for person detection in overhead images. If people are seen with their whole body in the area seen by the camera, the detection process is usually very successful. However, as seen in Fig. 1.11 a, the head of the person is not visible; other parts are visible. In this case, it appears that the detection process has failed. The reason for this is that people are labeled with their whole body during the image labeling stage.

Figure 1.12 shows the visual presentation of the test results obtained with the original Yolov4-Tiny and multi-scale Yolov4-Tiny algorithms. The visuals in the first column are the results obtained with the Yolov4-Tiny algorithm. The visuals in the second column belong to the multi-scale Yolov4-Tiny algorithm. When the results are examined, it seems that the multi-scale Yolov4-Tiny algorithm offers a better performance.

Table 1.2 summarizes the results for the five models. As can be seen from the values in the table in general, the greater the weight, the higher the success rate. While increasing the number of layers positively affects the success performance, it negatively affects the inference time. Yolov3's inference time is less than Yolov4's inference time. Yolov4-Tiny's inference time is lower than Yolov3-Tiny. It is observed that the FPS of the multi-scale YOLOv4-tiny is lower than the FPS value of YOLOv4-tiny, and it is considerably higher than the Yolov3 and Yolov4 algorithms. Yolov4 and Yolov3 algorithms seem to be the most successful models. These algorithms are followed by the multi-scale Yolov4-Tiny algorithm. The success of the multi-scale

Fig. 1.11 Visual results of the proposed method on Dataset 2

Yolov4-Tiny algorithm, which has a very low layer number compared to Yolov3, is very close to the Yolov3 algorithm. When the results are examined in terms of inference time and accuracy, it can be said that the proposed multi-scale Yolov4-Tiny algorithm has a good trade-off.

In addition, it was investigated whether there was a statistically significant difference between the proposed method and other yolo-based methods. In this context, p-value values for each pair of methods were calculated using the Wilcoxon signed-ranks test. The results of the analyzes performed at 0.05 significance level are presented in Table 1.3. A p-value of less than 0.05 (typically < 0.05) is statistically significant. This shows that there is strong evidence against the null hypothesis. It can be concluded from Table 1.3 that there are statistically significant differences between all method pairs.

1.5.1 Impact of the Different NMS Algorithms

At this stage, the effect of NMS algorithms on detection performance was investigated. Accordingly, the results of the experiments performed using the original NMS,

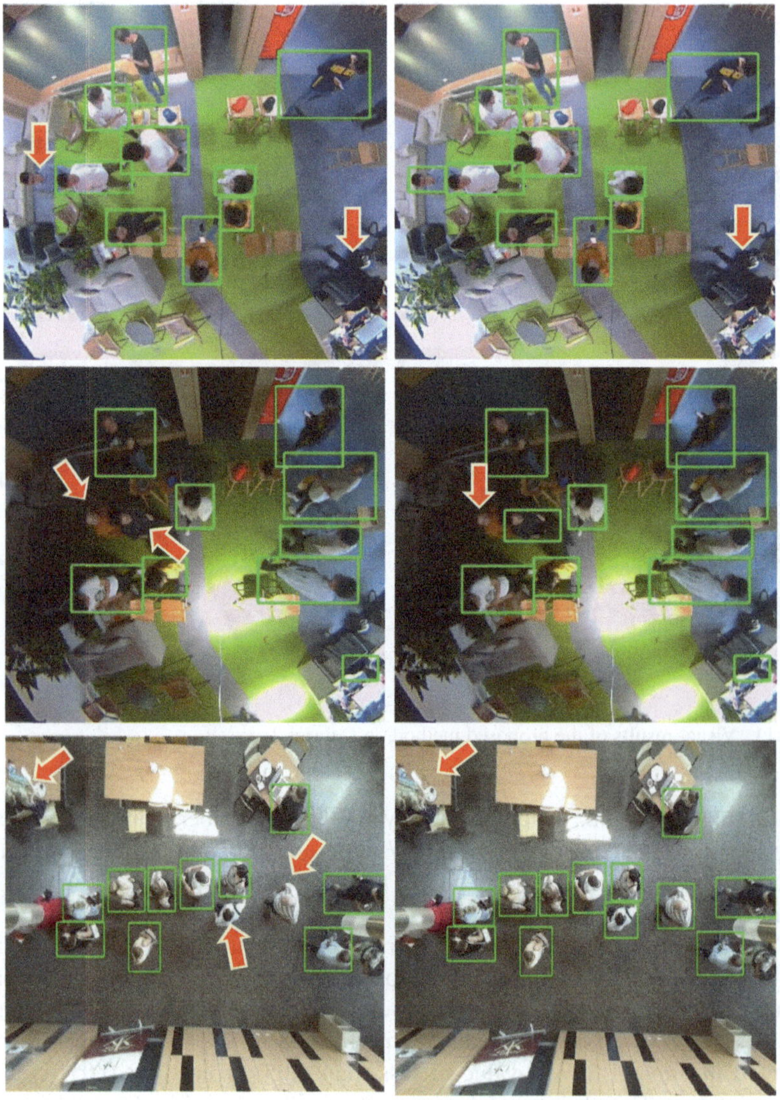

(a) (b)

Fig. 1.12 Comparison of **a** Yolov4-Tiny and **b** Multi-scale Yolov4-Tiny

soft-NMS and DIoU-NMS algorithms are presented in Fig. 1.13. When Fig. 1.13 is examined, better results were obtained with Soft-NMS compared to Original NMS. It is seen that these two algorithms show similar performance at 0.47 and 0.48 threshold values. It can be seen that the DIoU-NMS is better than the original NMS and Soft-NMS for each threshold. Overall, it seems that even the worst performance

Table 1.2 Comparative analysis of the proposed method and other Yolo-based methods

Model	P	R	F1	mAP (%)	FPS	Inference time / ms	Weights / MB
Yolov3	0.96	0.91	0.93	**95.42**	86	11.63	235
Yolov3-Tiny	0.72	0.83	0.77	75.20	**392**	**2.55**	33.1
Yolov4	0.98	0.93	0.95	**96.50**	80	12.50	244
Yolov4-Tiny	0.92	0.88	0.90	91.26	**405**	**2.47**	22.4
Multi-Scale Yolov4-Tiny	0.94	0.90	0.92	**94.85**	**380**	**2.63**	26.2

Table 1.3 The results of Wilcoxon signed-rank test for each pair of detection models

	Yolov3	Yolov3-Tiny	Yolov4	Yolov4-Tiny	Multi-Scale Yolov4-Tiny
Yolov3	–	0.000	0.047	0.000	0.000
Yolov3-Tiny	0.000	–	0.000	0.033	0.000
Yolov4	0.045	0.000	–	0.000	0.000
Yolov4-Tiny	0.000	0.047	0.000	–	0.033
Multi-Scale Yolov4-Tiny	0.000	0.000	0.000	0.033	–

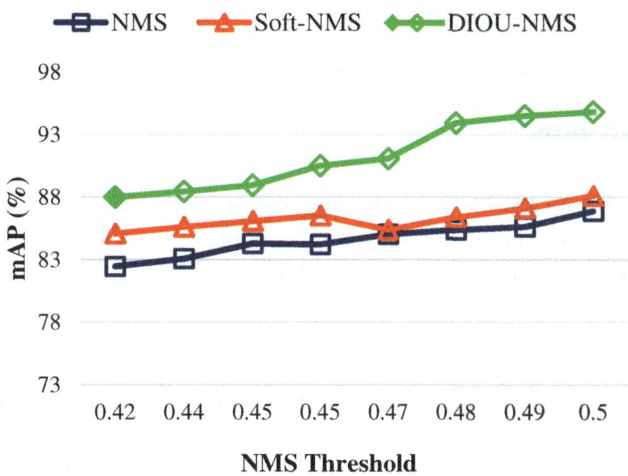

Fig. 1.13 Impact of the different NMS algorithms on people detection

of DIoU-NMS is comparable or better than the best performance of the original NMS and Soft-NMS algorithms.

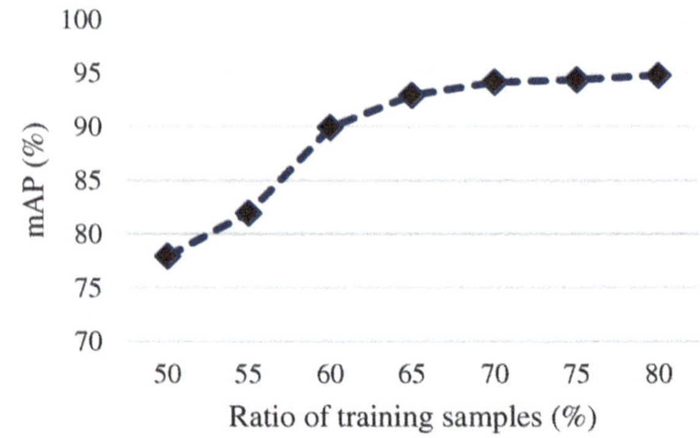

Fig. 1.14 The mAP scores obtained at different training samples rates

1.5.2 Impact of Training Dataset Size

Figure 1.14 shows the comparison of the ratios of different training samples and the different algorithms. It is seen that the OA of each algorithm increases with increasing training samples. This is because more information can be obtained as the number of training samples increases, thereby increasing the distinguishability of the algorithm and improving classification accuracy. When the training data exceeds 70% of the size of the dataset, the speed of performance improvement gradually slows down and tends to saturate. For a small number of training samples, the proposed algorithm still has good classification performance.

1.6 Conclusions

In this study, a deep learning-based method is proposed to detecting people from overhead images. The proposed method is an improved version of the Yolov4-Tiny algorithm. This method, called multi-scale Yolov4-Tiny, has four detection scales, unlike the original Yolov4-Tiny algorithm. In this way, it is possible to detect people who appear small in a camera area. In addition, the effects of the NMS algorithm used in the post-processing stage, which is an important stage of the detection process, on the performance were investigated. In this context, comparative analyzes were made with three different NMS algorithms. As a result of the experiments performed on data consisting of difficult scenarios such as occlusion, distance, closed light and wearing a hat, 94.85% mAP value was obtained with the proposed method. It has been observed that the method, which reaches high FPS values, is suitable for real-time applications. In addition, comparative analysis was done with different Yolo-based algorithms. As a result of the analysis, it has been seen that the multi-scale Yolov4-Tiny algorithm is a good option in terms of a trade- off based on computation time and accuracy.

In future studies, it is aimed to make experiments on a more comprehensive dataset. In addition, it is planned to implement tensor based optimizations (TensorRT) in order for the developed method to work more effectively in embedded systems.

Conflicts of Interest The authors declare that they have no conflicts of interest.

References

1. I. Ahmed, M. Ahmad, A. Ahmad, G. Jeon, Top view multiple people tracking by detection using deep SORT and YOLOv3 with transfer learning: within 5G infrastructure. Int. J. Mach. Learn. Cybern. 1–15 (2020)
2. C. Mohan, H.K. Verma, Direction and distance sensors and sensing system for elderly people. Mater. Today-Proc. **34**, 667–674 (2021)
3. R. Grimming, B. McIntosh, A. Mahalanobis, R.G. Driggers, LWIR sensor parameters for deep learning object detectors. OSA Continuum **4**(2), 529–541 (2021)
4. C.A. Luna, C. Losada-Gutiérrez, D. Fuentes-Jiménez, M. Mazo, Fast heuristic method to detect people in frontal depth images. Expert Syst. Appl. **168**, 114483 (2021)
5. M. Peker, B. Inci, E. Musaoglu, H. Cobanoglu, N. Kocakir, O. Karademir, Performance analysis of deep learning architectures on embedded devices for people counting system. Paper presented at the 5th International Mediterranean Science and Engineering Congress, Antalya, 20–22 April 2020 (2020)
6. S. Zhou, M. Ke, J. Qiu, J. Wang, A survey of multi-object video tracking algorithms, in *International Conference on Applications and Techniques in Cyber Security and Intelligence*. (Springer, New York, 2018), pp. 351–369
7. P. Li, D. Wang, L. Wang, H. Lu, Deep visual tracking: review and experimental comparison. Pattern Recogn **76**, 323–338 (2018)
8. L. Anuj, Krishna MG (2017) Multiple camera based multiple object tracking under occlusion: a survey. Paper presented at the International Conference on Innovative Mechanisms for Industry Applications (ICIMIA) **21–23**, 432–437 (Feb 2017)
9. R. Nakatani, D. Kouno, K. Shimada, T. Endo, A person identification method using a top-view head image from an overhead camera. J Adv Comput Intell Intell Inform **16**(6), 696–703 (2012)
10. M. Ahmad, I. Ahmed, K. Ullah, I. Khan, A. Adnan, Robust background subtraction based persons counting from overhead view. Paper presented at the 9th IEEE Annual Ubiquitous Computing, Electronics Mobile Communication Conference (UEMCON), 8–10 Nov 2018, pp 746–752 (2018)
11. L. Del Pizzo, P. Foggia, A. Greco, G. Percannella, M. Vento, A versatile and effective method for counting people on either RGB or depth overhead cameras. Paper presented at the IEEE International Conference on Multimedia & Expo Workshops (ICMEW), 29 June-3 July 2015, pp. 1–6 (2015)
12. S. Mukherjee, B. Saha, I. Jamal, R. Leclerc, N. Ray, A novel framework for automatic passenger counting. Paper presented at theIEEE International Conference on Image Processing (ICIP), 11–14 Sept. 2011, pp. 2969–2972 (2011)
13. S. Sun, N. Akhtar, H. Song, C. Zhang, J. Li, A. Mian, Benchmark data and method for real-time people counting in cluttered scenes using depth sensors. IEEE Trans Intell Transp Syst **20**(10), 3599–3612 (2019)
14. V. Carletti, L. Del Pizzo, G. Percannella, M. Vento, An efficient and effective method for people detection from top-view depth cameras. Paper Presented at the IEEE International Conference on Advanced Video and Signal Based Surveillance (AVSS), 29 Aug-1 Sept 2017, pp. 1–6 (2017)
15. M.S. Kristoffersen, J.V. Dueholm, R. Gade, T.B. Moeslund, Pedestrian counting with occlusion handling using stereo thermal cameras. Sensors **16**(1), 62 (2016)

16. I. Ahmed, A. Ahmad, F. Piccialli, A.K. Sangaiah, G. Jeon, A robust features-based person tracker for overhead views in industrial environment. IEEE Internet Things J **5**(3), 1598–1605 (2017)
17. I. Ahmed, A. Adnan, A robust algorithm for detecting people in overhead views. Cluster Comput **21**(1), 633–654 (2018)
18. I. Ahmed, M. Ahmad, A. Adnan, A. Ahmad, M. Khan, Person detector for different overhead views using machine learning. Int J Mach Learn Cybern **10**, 2657–2668 (2019)
19. M. Ahmad, I. Ahmed, A. Adnan, Overhead view person detection using YOLO. Paper Presented at the 10th Annual Ubiquitous Computing, Electronics & Mobile Communication Conference (UEMCON), 10–12 Oct. 2019, pp. 627–633 (2019)
20. M. Ahmad, I. Ahmed, K. Ullah, M. Ahmad, A deep neural network approach for top view people detection and counting. Paper presented at the 10th annual ubiquitous computing, electronics & mobile communication conference (UEMCON), 10–12 Oct 2019, pp. 1082–1088 (2019)
21. I. Ahmed, S. Din, G. Jeon, F. Piccialli, Exploring deep learning models for overhead view multiple object detection. IEEE Internet of Things J **7**(7), 5737–5744 (2019)
22. A. Musaev et al., Towards in-store multi-person tracking using head detection and track heatmaps (2020). arXiv preprint arXiv:2005.08009
23. Top View Multi Person Tracking Dataset, Ukrainian Catholic University, Rockville (2020). https://github.com/ucuapps/top-view-multi-person-tracking
24. M. Rezaei, M. Azarmi, DeepSOCIAL: social distancing monitoring and infection risk assessment in COVID-19 pandemic (2020). arXiv preprint arXiv:2008.11672
25. K. Simonyan, A. Zisserman, Very deep convolutional networks for large-scale image recognition (2014). arXiv preprint arXiv:1409.15561-13
26. K. He et al., Deep residual learning for image recognition. Paper presented at the IEEE Conference on Computer Vision and Pattern Recognition, 27–30 June 2016, pp. 770–778 (2016)
27. C.Y. Wang et al., CSPNet: a new backbone that can enhance learning capability of CNN. Paper Presented at the IEEE/CVF Conference on Computer Vision and Pattern Recognition Workshops, 14–19 June, 2020, pp. 390–391 (2020)
28. A. Sharifi et al., DeepHAZMAT: hazardous materials sign detection and segmentation with restricted computational resources (2020). Available at SSRN: https://ssrn.com/abstract=3649600
29. W. Liu et al., SSD: single shot multibox detector. Paper Presented at the European Conference on Computer Vision. Springer, pp. 21–37 (2016)
30. J. Redmon et al., You only look once: unified, real-time object detection. Paper Presented at the IEEE Conference on Computer Vision and Pattern Recognition. **2016**, 779–788 (2016)
31. J. Redmon, A. Farhadi, YOLOv3: an incremental improvement. Comput Vis Pattern Recognit (cs.CV) (2018). arXiv:1804.02767
32. A. Bochkovskiy, C.-Y. Wang, H.-Y.M. Liao, YOLOv4: Optimal speed and accuracy of object detection (2020). arXiv Comput. Vis. Patter Recognit. doi: arXiv: 2004.10934
33. Z. Jiang, L. Zhao, S. Li, Y. Jia, Real-time object detection method based on improved YOLOv4-tiny (2020). arXiv:2011.04244[cs]
34. Z. Zheng, P. Wang, W. Liu, J. Li, R. Ye, D. Ren, Distance-IoU loss: Faster and better learning for bounding box regression. Proceedings of the AAAI Conference on Artificial Intelligence **34**(07), 12993–13000 (2020)
35. N. Bodla, B. Singh, R. Chellappa, Davis LS (2017) Soft-NMS–improving object detection with one line of code. Paper Presented at the IEEE International Conference on Computer Vision **22–29**, 5561–5569 (Oct 2017)
36. F. Wilcoxon, Individual comparisons by ranking methods. Biom. Bull. **1**, 80–83 (1945)

Chapter 2
Machine Learning Techniques for Grocery Sales Forecasting by Analyzing Historical Data

Nilufa Yeasmin, Saman Hassanzadeh Amin, and Babak Mohamadpour Tosarkani

Abstract Product sales forecasting is a key aspect of purchasing management. Adoption of predictive analytics helps estimate market demand and determine inventory stock levels, which have been ongoing challenges, especially in the supermarkets and grocery stores. The total expected profit is reduced if a required level of products is not available. As a result, sales forecasting for goods has a significant impact to minimize the total costs associated with the lost opportunity. The purpose of this study is to create a forecasting model using machine learning algorithms, to get accurate forecasts for product sales. In this regard, several regression models (i.e., random forest, linear regression, decision tree) are employed and then, their results are discussed in detail using a grocery store's data set. Furthermore, the feed forward artificial neural network is utilized to optimize the results.

Keywords Decision tree · Random forest · Linear regression · Artificial Neural Network (ANN)

2.1 Introduction

The performance of entities (e.g., suppliers and retailers), incorporating to supply chain networks; depends on the sales forecasting [9]. The inaccurate prediction of grocery sales may result in either under stocking or over stocking situations in which stores incur additional costs. For example, if a particular item becomes out-of-stock, sellers are not able to make profit in case of market demand. Furthermore, if this

N. Yeasmin · S. H. Amin (✉)
Department of Mechanical and Industrial Engineering, Ryerson University, Toronto, ON, Canada
e-mail: saman.amin@ryerson.ca

N. Yeasmin
e-mail: nilufa.yeasmin@ryerson.ca

B. M. Tosarkani
School of Engineering, University of British Columbia, Kelowna, BC, Canada
e-mail: babak.tosarkani@ubc.ca

S. L. Fernandes and T. K. Sharma (eds.), *Artificial Intelligence in Industrial Applications*, Learning and Analytics in Intelligent Systems 25,
https://doi.org/10.1007/978-3-030-85383-9_2

under stocking situation occurs regularly, customers become disappointed to return for their future purchases [6].

In this research, a grocery data set, acquired from the "Kaggle" website [8], is utilized to forecast product sales based on the items, stores, transactions, and other dependent variables such as holidays and oil prices. This is a competition called "Corporación Favorita Grocery Sales Forecasting" by "Kaggle" applied to predict the required levels of grocery items to fulfill market demand. To this aim, we intend to analyze the data as a supervised learning problem. In order to forecast the sales, different regression models (e.g., random forest, linear regression, and decision tree) are considered. Then, a feed forward Artificial Neural Network (ANN) is developed to optimize the results.

2.2 Literature Review

The primary focus of this study is to develop a model, using machine learning algorithms to forecast the sales. In this section, the most relevant papers related to sales forecasting using different regression models and ANN are reviewed. ANN has been effectively applied in many sectors. In the field of supply chain management, ANN is widely applied in different potential areas such as forecasting, decision-making, logistics optimization, and enterprise resource planning. Cottrell et al. [7] described the problem of network arrangement for product sales forecasting based on time series. They integrated statistical techniques (linear and nonlinear time series analyses) with connection approach. Yip et al. [22] investigated the sales data of "Glynwed Foundry Products Limited" of the United Kingdom from January 1986 to October 1992. The main objectives of their study were to realize suitable permutations of inputs and to predict sales up to six months in advance. Furthermore, they compared the ANN results with the results of time series smoothing method. They also compared the test outcomes of time series smoothing methods of forecasting by applying some methods. Zhang et al. [24] provided additional widespread review of the present standing regarding forecasting with ANNs. They described several fields in which ANN has been applied for sales forecasting. The comparative performance of ANNs was discussed over conventional statistical methods in their paper.

The characteristics of the data sets, incorporating to a supply chain, should be taken into account prior to the adoption of machine learning algorithms. Ali et al. [3] conducted a detailed exploratory analysis of the data set to develop a model using linear regression, support vector regression, and regression tree. Khashei and Bijari [13] developed an ANN model to generate data using an Autoregressive Integrated Moving Average (ARIMA) model to forecast the forthcoming significance of time series method.

Lee et al. [15] proposed an Enhanced Cluster and Forecast Model (ECFM) for sales forecasting of different foods. They also developed a Self-Organization Map (SOM), and a Radial Basis Function (RBF) ANNs. The proposed model was developed based on the sales of 6-month records of a grocery shop in Taiwan. The result was computed based on a Fuzzy Neural Network (FNN) model, and a Cluster Forecast Model (CFM). Lu et al. [19] developed a Multivariate Adaptive Regression Splines (MARS) model related to sales forecasting for computer wholesalers. In the airline industry, decision-makers usually adopt Neural Networks in forecasting airline passenger loads due to the high accuracy. Kandananond [11] conducted two types of data mining methods (i.e., Support Vector Machine (SVM) and ANN) to predict the market demand. They applied training data related to demand of 6 individual items from a consumer goods company in Thailand. The computational results indicated that SVM had good prediction quality based on the Mean Absolute Percentage Error (MAPE) with the high boundary division of MAPE. Ahmad et al. [1] proposed an electrical energy forecasting method through the application of machine learning methods such as SVM and ANN.

Leksakul et al. [16] established a forecasting model based on off-season supply by using methods of fuzzy SVM and fuzzy ANN. Furthermore, they developed a grid search for the SVM to discover finest model and applied real-world data to estimate the proficiency of the executed algorithms. Liu [17] proposed a neural network model and linear regression to implement a forecasting analysis for a bike market demand. A learning algorithm Back Propagation (BP) NN was applied while the number of hidden layers were assumed to be 1. Patel et al. [20] proposed a model for moving stock exchange market and stock price for dashboards of Indian stock exchange markets. They developed several prediction models using machine learning algorithms such as SVM, ANN, random forest, and Naive-Bayes. Kochak and Sharma [14] developed forward and BP NN in MATLAB to predict sales of fuel filter based on the historical sales data. Bhadouria and Jayant [4] investigated a gear manufacturing company, and developed three ANN models with TANSIGMOID, LINEAR and LOGSIGMOID transfer function using MATLAB for the forecast of market demand related to different items. Wijnhoven and Plant [21] analyzed 500,000 posts of social media for eleven car models in the Dutch market. They used linear regression model to predict the car sales. The results suggested that Google Trends data has higher predictive impact on car sales in comparison with other social media. Loureiro et al. [19] developed a deep learning methodology to forecast the trades for the future values of products in the fashion industry. Kaya and Turkyilmaz [12] employed an intermittent demand predicting model through the application of machine learning techniques (e.g., ANN and SVM). In another important paper, Huang et al. [10] proposed operational approaches for forecasting retail product sales. Their method focused on the efficient consumption of available promotional evidence. A review of relevant studies is provided in Table 2.1.

2.3 Exploratory Data Analysis

Two different types of analyses are conducted since the considered data sets in this study include text and numeric fields.

Table 2.1 Review of some studies utilizing linear regression, decision tree, random forest, and ANN

Authors	Method(s)*	Data type
Cottrell et al. [7]	ANN with activation linear and Tanh and performance measure by residual variance	Yearly sunspots
Yip et al. [22]	ANN	Store product sales data
Zhang et al. [24]	ANN over traditional statistical methods	Grocery store sales data
Ali et al. [3]	Linear regression, support vector regression, regression tree	SKU- Grocery store sales data
Khashei and Bijari [13]	ANN, ARIMA	Dataset Wolf's sunspot, Canadian lynx data and British pound/United States dollars exchange amount
Kandananond [11]	ANN, SVM	Different types of product category
Lee et al. [15]	Self-Organization Map (SOM), ANN, and Radial Basis Function (RBF)	Daily fresh foods sales data of chained shop in Taiwan
Lu et al. [19]	ANN with multivariate adaptive regressions plines	Computer wholesalers
Ahmad et al. [1]	SVM, ANN	Energetic materials
Kochak and Sharma [14]	Forward and back propagation NN	Sales data
Leksakul et al. [16]	Fuzzy support vector regression, ANN, SVM, and Fuzzy NN	Longan dataset
Liu [17]	ANN, Linear regression	Bike market data
Patel et al. [20]	ANN, SVM, random forest and Naive-Bayes (NV)	Stock trading data
Bhadouria and Jayant [4]	ANN models with TANSIGMOID, LINEAR, and LOGSIGMOID	Gear manufacturing data
Wijnhoven and Plant [21]	Linear regression	Social media posts for eleven car models
Loureiro et al. [19]	DT, random forest, SVM, ANN and linear regression	Historical sales data
Kaya and Turkyilmaz [12]	ANN, SVM	Generate artificial demand data, Intermittent package of R
Huang et al. [10]	Least absolute shrinkage and selection operator (LASSO) procedure and Autoregressive Distributive Lag (ADL)	Retail data, Information Resources, Inc. (IRI) company

(continued)

Table 2.1 (continued)

Authors	Method(s)*	Data type
Our study	Linear regression, decision tree, random forest, ANN	Grocery store items

*Artificial Neural Networks (ANN), Support Vector Machine (SVM), Self-Organization Map (SOM), Naive-bayes (NV), Neural Network (NN)

2.3.1 Data Source and Data Files

In this study, the data sets are adopted from Kaggle website (i.e., open source). They deliver the information about grocery sales stores forecasting, including: (i) The Stores.csv file contains the details information about the stores (e.g., location and type). (ii) The Items.csv file describes item metadata, such as class and perishability. The perishable items have a higher scoring weight than the others. (iii) The Transactions.csv file counts the total sales covenant for data of training. (iv) The Oil.csv file shows the information about daily oil price, which is very important since Ecuador has the oil-dependent economy. (v) The Holidays_events.csv file carries relevant information about sales. People usually purchase their necessary goods on holidays. Therefore, sales increase during such periods. Figure 2.1 describes an overview of the individual data files for grocery sales forecasting.

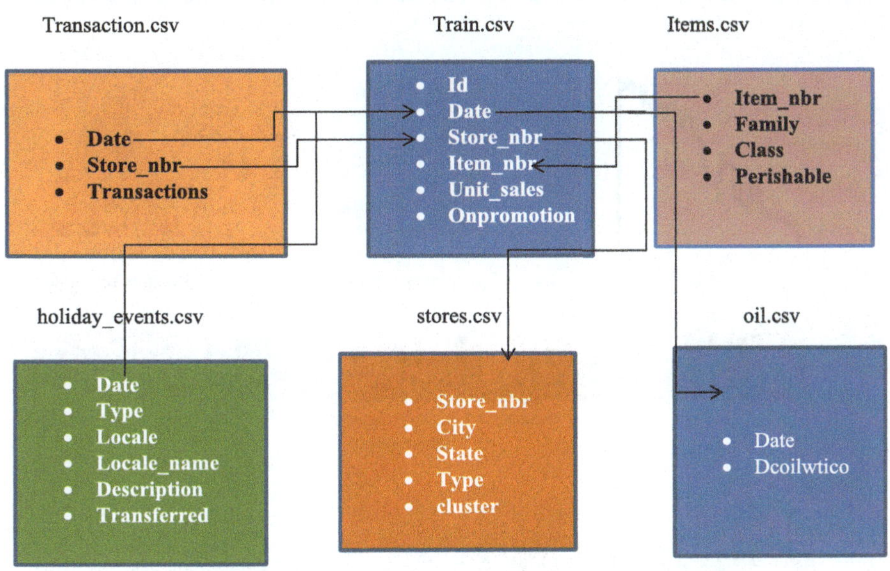

Fig. 2.1 Overview of individual data files

2.3.2 *Data Analysis*

To analyze the data sets, some required python libraries (e.g., pandas, numpy, matplotlib, and seaborne) are imported. Then, the data sets are loaded. The distinct feature distribution of each data set was plotted individually to demonstrate the overview of the data sets. Then, the impacts of individual features on the sales records are explored. Figure 2.2 illustrates the data distributions through some histograms (for numerical and date features) and bar plots (for categorical features).

Figure 2.3 indicates a sale per store type in which Store type A has the highest sales in comparison with the other store types. In addition, Fig. 2.4 displays that the most stores are located in the cities of Quito and Guayaquil.

Figure 2.5 illustrates that the states of Pichincha Santo and Guayas have the most Stores.

In the next phase of data analysis, the type of stores in every city is investigated. As shown in Fig. 2.6, every city may not have all types of stores.

There is a variety of events arranged during holidays. As a result, the sales increase during such periods. Figure 2.7 demonstrates the type of events occurring in every city.

Figure 2.8 shows the time series of accumulative transactions at each store. The median values of all transactions are calculated for the stores. The red line describes the information about the smoothing fit.

Figure 2.9 demonstrates that there are longstanding changes of oil price with an obvious fall in the second half of 2014. The frequent downward trends are visible in

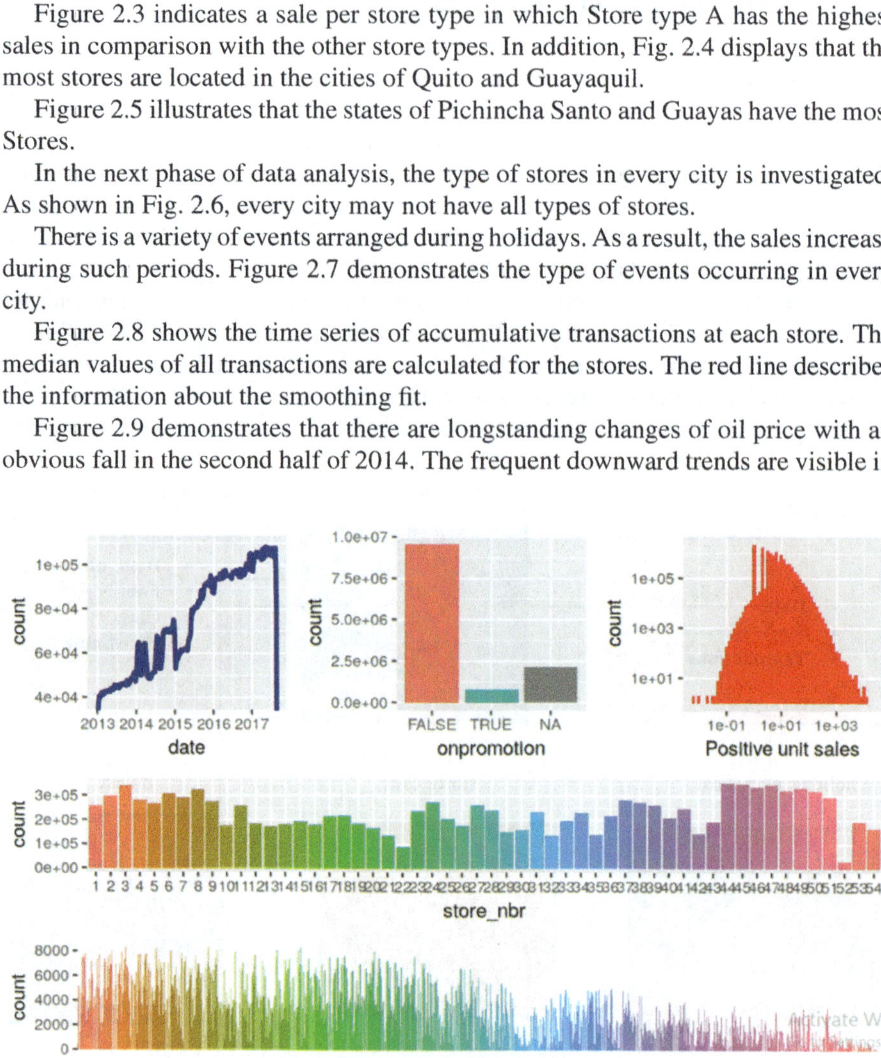

Fig. 2.2 Overview of training data set

Out[37]: <matplotlib.axes._subplots.AxesSubplot at 0xb1815ab828>

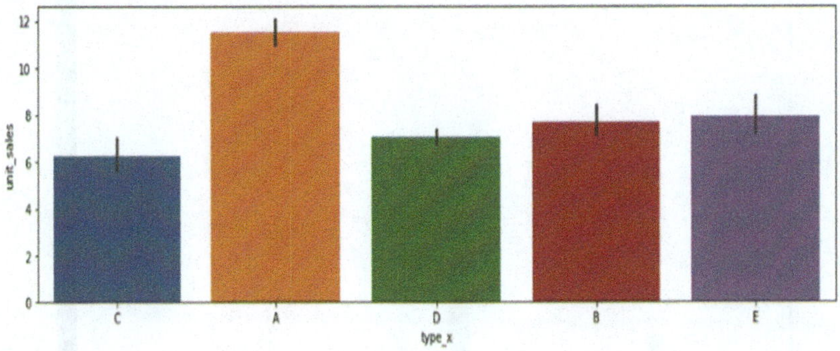

Fig. 2.3 Sales per store type

Out[44]: <matplotlib.axes._subplots.AxesSubplot at 0xb18202c828>

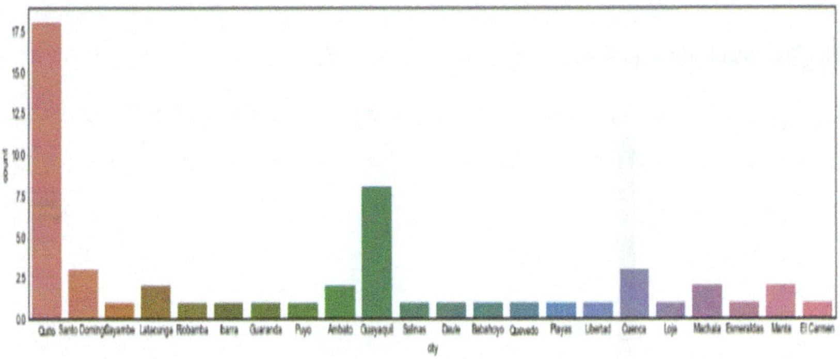

Fig. 2.4 Maximum stores in cities

Out[49]: <matplotlib.axes._subplots.AxesSubplot at 0xb182ded470>

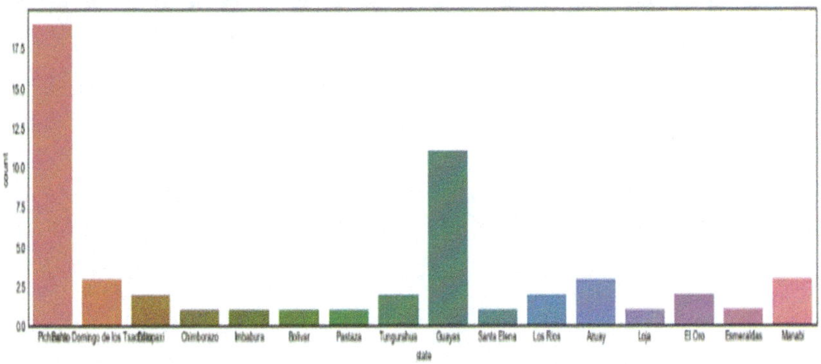

Fig. 2.5 Maximum stores in the states

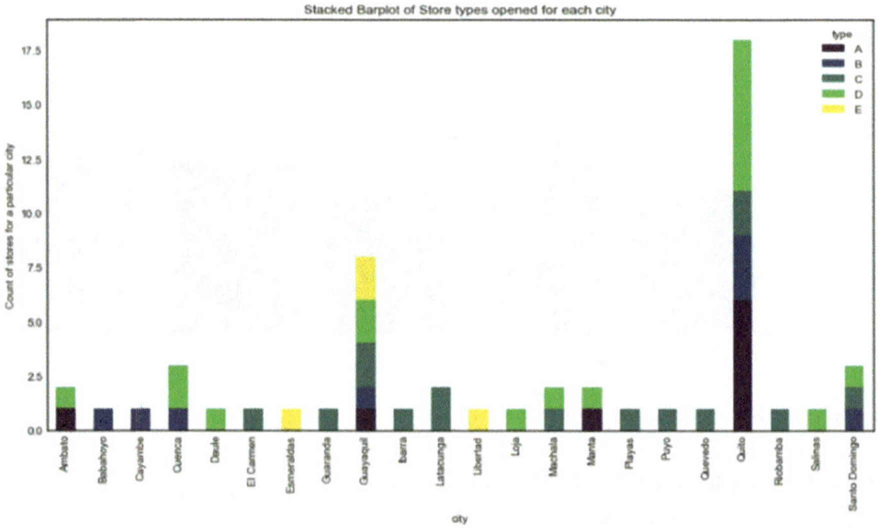

Fig. 2.6 Stacked bar plot indicating store types in every city

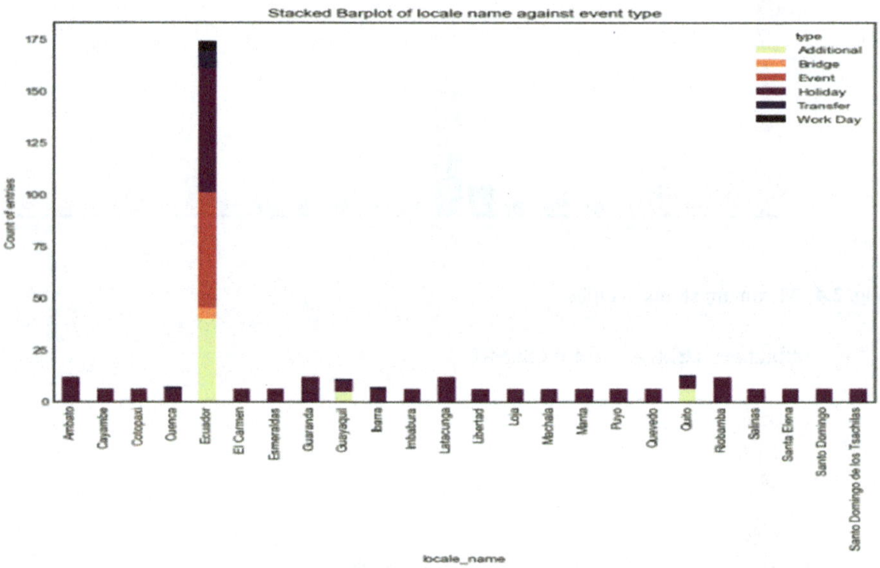

Fig. 2.7 Stacked bar plot of event type occurring in every city

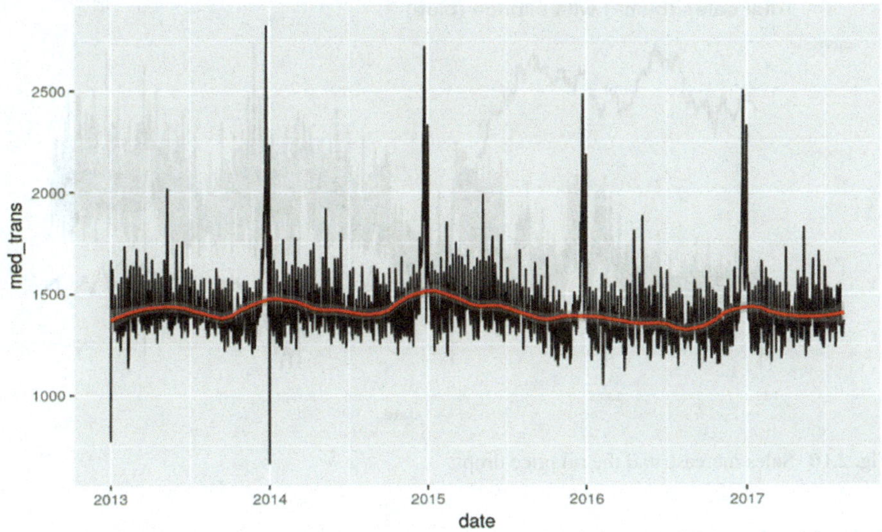

Fig. 2.8 Cumulative transactions of per store

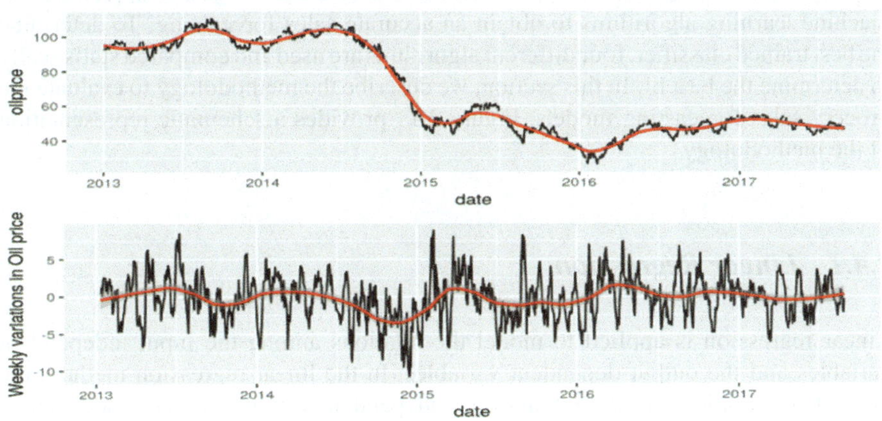

Fig. 2.9 Cumulative transactions of per store

the week-to-week variations. The strong downs and rises in the oil price may have a significant impact on the purchasing power of oil-dependent economy. The oil price over time along with its weekly fluctuations are plotted in Fig. 2.9.

Figure 2.10 indicates the relation between the oil price and the sales. The considerable oil price fall happened during the time of the second half of 2014, which led to the reduction of sale numbers in the first half of 2015. The oil price was comparatively unwavering at the late 2017.

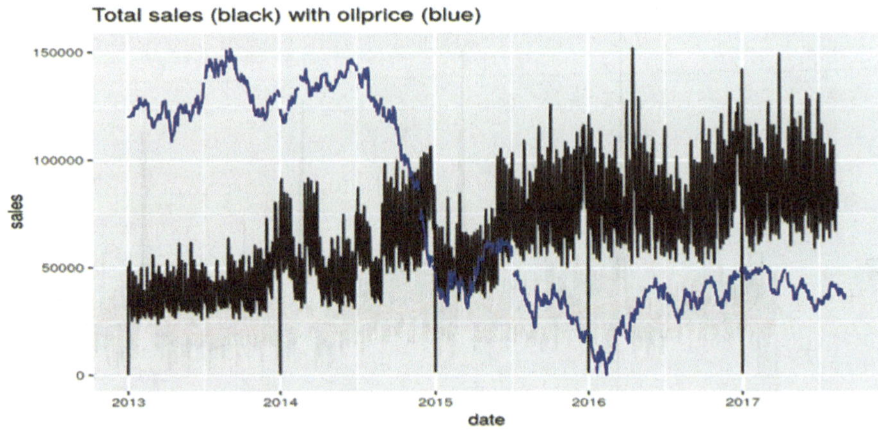

Fig. 2.10 Sales increase and the oil price drops

2.4 Methodology

The main purpose of this study is to develop a model through the application of machine learning algorithms to obtain an accurate sales forecasting. To determine the best trained classifier, four different algorithms are used and compared statistically to determine the best fit. In this section, we describe the methodology to evaluate the grocery sales forecasting models. Figure 2.11 provides a schematic representation of the methodology.

2.4.1 Linear Regression

Linear regression is applied to model the relations among the input independent variables and the output dependent variables. In the linear regression method, the dependent variable is incessant, and the independent variables can be incessant or discrete. If input variables and output variables are shown by X and Y, respectively, then the linear regression approach creates a relation between X and Y by using a finest fit straight line (i.e., the regression line). In this regard, linear regression algorithm estimates the values of the coefficients through the application of several techniques (e.g., ordinary least squares, gradient descent, regularization) (Kumari and Yadav 2018).

The ordinary least squares method is used when there are several inputs to evaluate the standards of coefficients. Gradient descent is a repetitive optimization technique used to find the local minimum of a function. Gradient descent is applied based on the random values for each iteration of coefficient (Kumari and Yadav 2018). The sum of the squared errors is considered for both input and output values. Regularization is the delay of training linear model which reduces the sum of the squared error of training

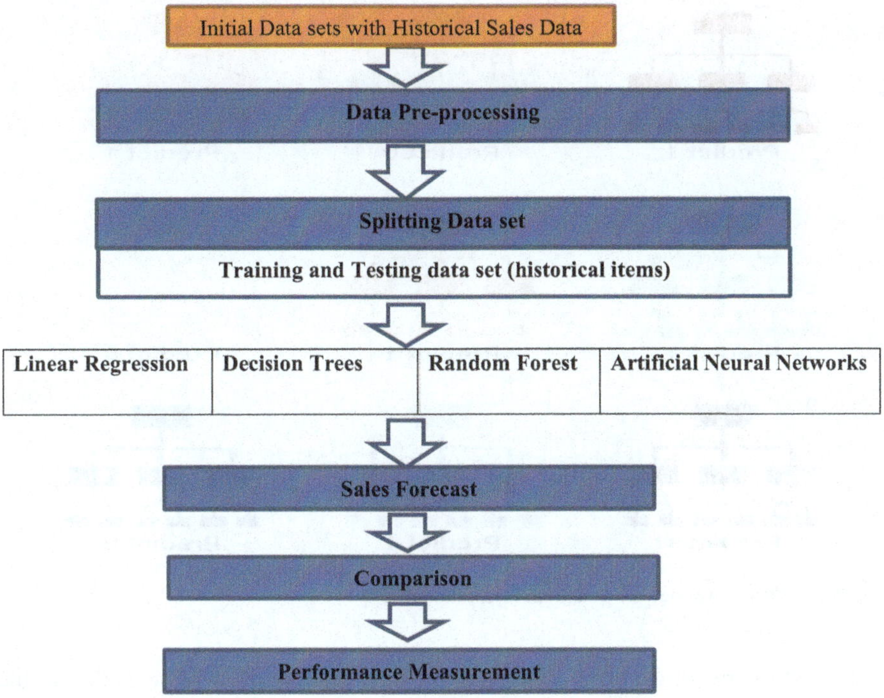

Fig. 2.11 Schematic representation of the methodology

basis (consuming ordinary least squares), and also decreases the complication of the model (absolute size of sum of all coefficients in the model). R-squared (R^2) denotes the amount of the variance for the dependent variable described by the independent variable(s).

2.4.2 Decision Trees

A decision tree is a decision support approach, tree-like model, representing decisions and their possible consequences [19]. This approach has a flow-chart-like structure including multiple nodes (i.e., attributes), branches (i.e., outcomes of the experiment), and leaves (i.e., original outputs).

2.4.3 Random Forest

A random forest technique consists of a large number of separate decision trees which work as same as ensemble models. In the random forest model, individual trees are

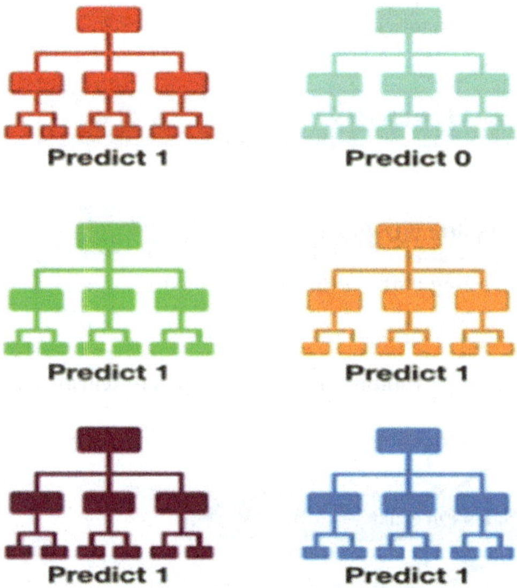

Fig. 2.12 Visual overview of a random forest model [23]

employed to predict the class containing the maximum number of votes that would be our model's prediction [2]. Figure 2.12 provides a clear concept about random forest technique, including six predict 1 tree, and three predict 0 tree. Therefore, the prediction result is 1.

2.4.4 Artificial Neural Networks

ANN is an information processing technique of machine learning that is actually motivated based on human nervous structure (e.g., the information processing system of brain) (Kaya and Turkyilmaz [12]). The network architecture of ANN consists of the input layer, bias, and output layer. ANN is also known as an information processing model that is motivated by the way of human brain structure. In machine learning, perception is an algorithm that actually works for the supervised learning. Figure 2.13 denotes the basic NN motivated by a biological brain structure (i.e., concept of perception). In this single layer NN, x_0, x_1, x_2, $x_3 \ldots x_n$ denote several inputs of the network for a single observation in which every input is multiplied by weight (e.g., $W_{0,} W_1$, W_2, W_n). b can be a bias unit, or bias is an activation of the neural network. In the other words, the inputs are multiplied with the weights ($x_1 w_1 + x_2 w_2 + x_3 w_3 \ldots\ldots x_n w_n = \sum x_i w_i$). Then, the activation function ($\sum x_i w_i$) is applied. There are many types of activations in ANN. In this study, Rectified Linear Units (RELU) activation is employed. RELU is an activation function in which the limits start from zero to infinity ($0, \infty$).

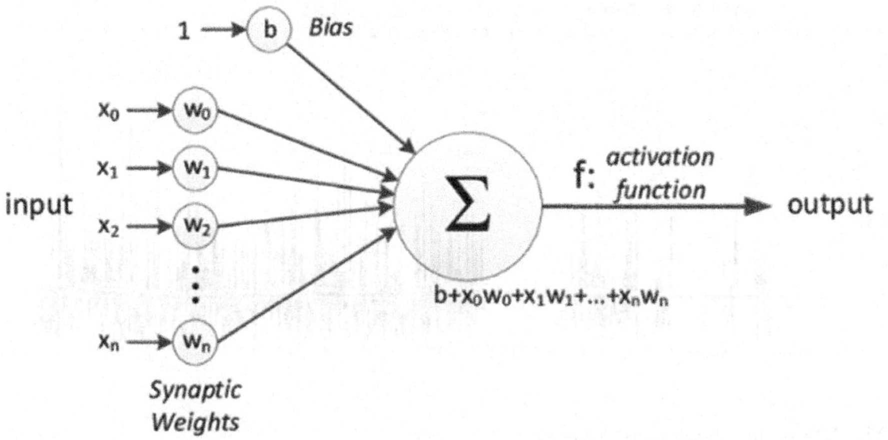

Fig. 2.13 Single layer neural network or perception [5].

2.5 Experimental Results and Analysis

Sales forecasting is a method for estimating the future demand of corporation's sales based on the historical data. To avoid risk and to minimize the loss of companies, the sales prediction model is essential. An accurate sales forecasting can help avoid unexpected cash drift difficulties and can increase the overall profit. This section provides information about the experimental results. For the first time experiment, time-series forecasting is applied since we have a large number of time series data. The relations between the various time series parameters (e.g., type of item, location of store) are estimated to improve the prediction. Figure 2.14 shows the information about the individual item of every store in different year. The plot title contains the information about the item classification (class and family) as well as the store type

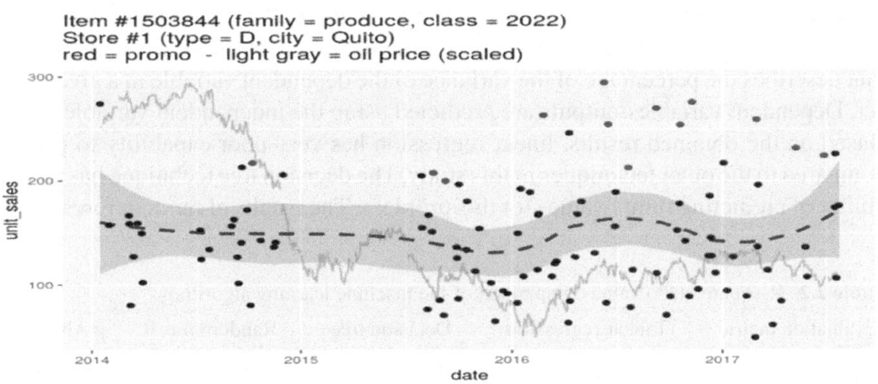

Fig. 2.14 Individual item of individual store

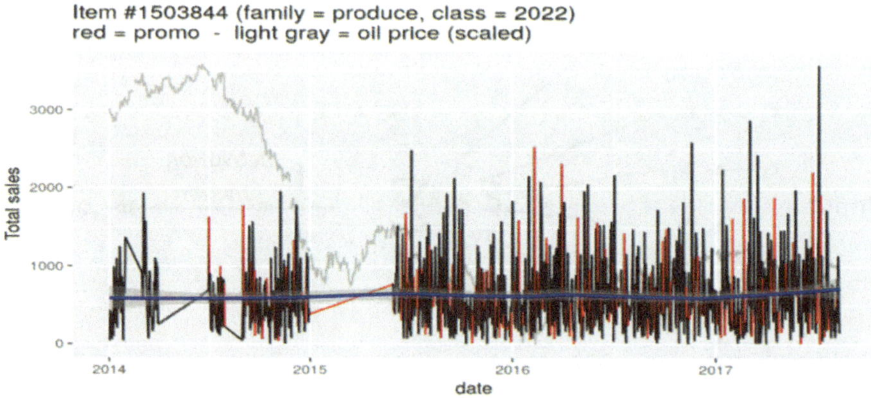

Fig. 2.15 Items on sales based on year

and the city. The red circles indicate the items that are promotional. The dashed blue line is a smoothing fit with the corresponding gray confidence range. This allows us to judge based on long-term variation. The light-gray curve in the background is the oil price during this time range, and the y-axis represents the unit sales.

Figure 2.15 demonstrates the individual item of individual store. We used a connected line instead of the data points. It emphasizes the short-term variations and the gaps in the information for this particular product. The red flag is a set if one of the stores has a promotion for the specific item.

In the second experiment, it is aimed to forecast the product sales based on the items, stores, transaction, and other dependent variables (e.g., holidays and oil prices). The data sets are analyzed as a supervised learning problem. In order to forecast the sales, we have compared the different regression models including decision tree, linear regression, and random forest. Then, the multi-layer perception and feed forward ANN are employed to optimize the results.

Table 2.2 describes the results of the experiments and the performance of the employed techniques. The evaluation metric *R-squared* (R^2) is utilized to illustrate a statistical measure as an accuracy measure. In this regard, R^2 is a statistical measure that describes the percentage of the variance of the dependent variable in a given data set. Dependent variables/outputs are predicted using the independent variable/input. Based on the obtained results, linear regression has very poor capability to predict compared to the other techniques in this study. The decision tree technique has a good ability of predicting future values for this problem. The results of random forest show

Table 2.2 R-squared (R^2) value comparison of the machine learning algorithm

Evaluation metric	Linear regression	Decision tree	Random forest	ANN
R^2	0.366	0.705	0.801	0.836

that this method has high capability to forecast the sales in the future. Based on Table 2.2, ANN has the highest accuracy to predict the sales based on the historical data.

2.6 Conclusions and Future Work

Sales forecasting is very essential for inventory planning especially for grocery items. An accurate forecasting model can help companies, involved in a grocery supply chain to increase their total expected profit. The main objective of this study is to develop a forecasting model by using machine learning algorithms. The proposed model has been targeted to support accurate forecasting of future purchases. A grocery store's data set has been investigated to validate the use of the numerous machine learning algorithms. In this study, multiple methods have been compared. The methods have affected the products availability in the store at the right time. Based on the findings of this study, linear regression technique had poor capability to forecast the upcoming product sales. On the other hand, ANN technique had a very high accuracy to calculate the future product sales.

One future research area is improving the proposed model by incorporating other predictive variables in the proposed model. For example, Google trends can be considered to get and estimate the market demand and update the dependent and independent variables in the proposed model. Furthermore, Multivariate Hybrid Intelligent (MHI) forecasting models can be implemented to forecast sales and improve the accuracy of the models. Accordingly, MHI will be a great research path to discover actual intelligent models for forecasting the future sales.

References

1. A.S. Ahmad, M.Y. Hassan, M.P. Abdullah, H.A. Rahman, F. Hussin, H. Abdullah, R. Saidur, A review on applications of ANN and SVM for building electrical energy consumption forecasting. Renew. Sustain. Energy Rev. **33**, 102–109 (2014)
2. Ö. Akar, O. Güngör, Classification of multispectral images using Random Forest algorithm. J. Geodesy Geoinf. **1**(2), 105–112 (2012)
3. Ö.G. Ali, S. Sayın, T. Van Woensel, J. Fransoo, SKU demand forecasting in the presence of promotions. Expert Syst. Applic. **36**(10), 12340–12348 (2009)
4. S. Bhadouria, A. Jayant, Development of ANN models for demand forecasting. Am. J. Eng. Res **6**, 142–147 (2017)
5. N.S. Chauhan, KDnuggets Online Data Science Platform (2019). https://www.kdnuggets.com/2019/10/introduction-artificial-neural-networks.html. Accessed March 28, 2021
6. D. Corsten, T. Gruen, Desperately seeking shelf availability: an examination of the extent, the causes, and the efforts to address retail out-of-stocks. Int. J. Retail Distrib. Manag. **31**(12), 605–617 (2003)
7. M. Cottrell, B. Girard, Y. Girard, M. Mangeas, C. Muller, Neural modeling for time series: a statistical stepwise method for weight elimination. IEEE Trans. Neural Networks **6**(6), 1355–1364 (1995)

8. Favorita-grocery-sales, Open dataset from Kaggle (2019). https://www.kaggle.com/c/favorita-grocery-sales-forecasting/data. Accessed March 28, 2021
9. R. Fildes, S. Ma, S. Kolassa, Retail forecasting: research and practice. Int. J. Forecasting (In Press)
10. T. Huang, R. Fildes, D. Soopramanien, Forecasting retailer product sales in the presence of structural change. Eur. J. Oper. Res. **279**(2), 459–470 (2019)
11. K. Kandananond, Consumer product demand forecasting based on artificial neural network and support vector machine. World Acad. Sci. Eng. Technol. **63**, 372–375 (2012)
12. G.O. Kaya, A. Turkyilmaz, Intermittent demand forecasting using data mining techniques. Appl. Comp. Sci. 14(2) (2018)
13. M. Khashei, M. Bijari, An artificial neural network (p, d, q) model for timeseries forecasting. Expert Syst. Appl. **37**(1), 479–489 (2010)
14. A. Kochak, S. Sharma, Demand forecasting using neural network for supply chain management. Int. J. Mechan. Eng. Robot. Res. **4**(1), 96–104 (2015)
15. W.I. Lee, B.Y. Shih, C.Y. Chen, Retracted: A hybrid artificial intelligence sales-forecasting system in the convenience store industry. Human Fact. Ergon. Manufact. Serv. Ind. **22**(3), 188–196 (2012)
16. K. Leksakul, P. Holimchayachotikul, A. Sopadang, Forecast of off-season longan supply using fuzzy support vector regression and fuzzy artificial neural network. Comput. Elect. Agricul. **118**, 259–269 (2015)
17. H. Liu, Forecasting model of supply chain management based on neural network. In 2015 International Conference on Automation, Mechanical Control and Computational Engineering. Atlantis Press (2015, April)
18. A.L. Loureiro, V.L. Miguéis, L.F. da Silva, Exploring the use of deep neural networks for sales forecasting in fashion retail. Decis. Support Syst. **114**, 81–93 (2018)
19. C.J. Lu, T.S. Lee, C.M. Lian, Sales forecasting for computer wholesalers: a comparison of multivariate adaptive regression splines and artificial neural networks. Decis. Support Syst. **54**(1), 584–596 (2012)
20. J. Patel, S. Shah, P. Thakkar, K. Kotecha, Predicting stock and stock price index movement using trend deterministic data preparation and machine learning techniques. Expert Syst. Appl. **42**(1), 259–268 (2015)
21. F. Wijnhoven, O. Plant, Sentiment analysis and google trends data for predicting car sales. In ICIS (2017)
22. D.H. Yip, E.L. Hines, W.W Yu, Application of artificial neural networks in sales forecasting. In Proceedings of International Conference on Neural Networks (ICNN'97) (Vol. 4, pp. 2121–2124). IEEE (1997)
23. T. Yiu, Towards Data Science (2019). https://towardsdatascience.com/understanding-random-forest-58381e0602d2, Accessed March 28, 2021
24. G. Zhang, B.E. Patuwo, M.Y. Hu, Forecasting with artificial neural networks: The state of the art. Int. J. Forecast. **14**(1), 35–62 (1998)

Chapter 3
Medication Revelation Utilizing Neural Network

Virendra Singh Kushwah, Ashish Solanki, Bhavya Manoj Votavat, and Aman Jain

Abstract In the fields of drug discovery and development, machine learning algorithms have been used for the development of novel drug candidates. The methods for designing drug targets and novel drug discovery now routinely combine machine learning and deep learning algorithms to enhance the efficiency, efficacy, and quality of developed outputs. The generation and incorporation of big data, through technologies such as high-throughput screening and high through-put computational analysis of databases used for both lead and target discovery, has increased the reliability of the machine learning and deep learning incorporated Algorithms. In this work, the applications that produce promising results and methods will be reviewed. The use of virtual screening and encompassing online information has also been highlighted in developing lead synthesis pathways.

3.1 Introduction

Artificial Intelligence (AI) is broadly utilized in both industry and the scholarly world. AI (ML), a significant segment in AI, has been coordinated into numerous fields, like information age and investigation. Proteins are the most flexible natural particles, as they cover jobs going from underlying scaffolding to dynamic freight

V. S. Kushwah (✉) · A. Solanki · B. M. Votavat · A. Jain
VIT Bhopal University, Sehore, India
e-mail: virendra.kushwah@vitbhopal.ac.in

A. Solanki
e-mail: ashish.solanki2019@vitbhopal.ac.in

B. M. Votavat
e-mail: bhavya.manoj2019@vitbhopal.ac.in

A. Jain
e-mail: aman.jain2019@vitbhopal.ac.in

S. L. Fernandes and T. K. Sharma (eds.), *Artificial Intelligence in Industrial Applications*, Learning and Analytics in Intelligent Systems 25,
https://doi.org/10.1007/978-3-030-85383-9_3

transport. Malignant growth is, above all else, a sickness of the genome, brought about by ecological mutagens or because of inadequate DNA fix systems [11]. In particular, the most punctual malignancy drugs were fortunately found and worked through the restraint of cell division on a living being wide scale [1, 5].

The world's clinical network began to utilize an allopathic way to deal with treatment and recuperation in the 19th century. This change prompted the achievement of battling illnesses, however high medication costs followed, turning into medical care trouble. AtomNet is the main Deep Convolution Neural Network (DCNN) for subatomic restricting partiality forecasts [13, 15]. Compound get-together are portrayed by the spatial strategy and holding of various particles in space, notwithstanding, these atoms are general to each other. The model design compels its model design to use the spatial and fleeting construction of its area. The biochemical affiliations are relatively neighborhood, and should be exhibited by equivalently constrained AI plans [6, 9]. The model design reduces over-fitting and improves the revelation of generalizable highlights. Besides, with edge finders in DCNNs for pictures, the suitability of a marker for e.g., hydrogen holding or π-bond stacking, is invariant across the responsive field. These neighborhood biochemical association locator's may then be dynamically made into more convoluted features depicting the flighty and nonlinear wonder of sub-nuclear confinement.

As of late, Deep neural networks have appeared to out-perform irregular timberlands and SVMs for QSAR [3, 14] and ligand-based virtual screening [10]. The best per-shaping design for the Merck Molecular Activity Kaggle Challenge was to perform multiple tasks on a deep neural network (MT-DNN) [2]. They perform various tasks to prepare a single neural network with numerous yield neurons, every one of which anticipates the action of the info atom in an alternate examination. Be-cause particles are regularly tried in various measures, the MT-DNN design can join preparing proof among comparable expectation undertakings. That showed the MT-DNN procedure scales to huge biochemical information bases, for example, PubChem Bioassays and ChEMBL [6, 9, 15].

Ligand-based systems, including MT-DNN [2], go with a couple of limitations. In the first place, they are confined to centers for which significant proportions of prior data are presently available and, accordingly, can't make estimates for novel targets. All things considered, this makes a bewildering dynamic - these perceptive models offer the most help precisely for those targets which least require it. The de-irregularity of acknowledged unique ligands similarly makes it difficult to show that the association is "fitting for the right reasons"; relics in the arrangement data, similar to basic inclination, make it hard to appropriately assess precision and generalizability. Second, existing significant neural associations for ligand-based models [10]take nuclear fingerprints, for instance, ECFP, as data. Such data encoding limits the disclosure of features to plans of the pre-shown sub-nuclear developments portrayed during the fingerprinting cycle and discards the ability to discover emotional features [12]. Third, as the model is careless in regards to the evenhanded, the model can't explain which potential participations are left unfulfilled by a molecule. This limits the heading that could be given to help researchers to smooth out the molecule [21].

3.2 Material and Methods

We initially depict the development of the test benchmarks on which we test our framework. At that point, we portray our information encoding and the plan of our deep convolutional network.

3.2.1 ML Algorithms used in Drug Discovery

ML algorithms have significantly advanced drug discovery [12, 15]. Drug organizations have significantly profited by the usage of different ML calculations in drug revelation. ML calculations have been utilized to create different models for foreseeing synthetic, organic, and physical qualities of mixtures in drug revelation [8]. ML calculations can be consolidated in all steps of the cycle of medication disclosure. For instance, ML calculations have been utilized to track down another utilization of medications, foresee drug-protein collaborations, find drug viability, guarantee security biomarkers, and enhance the bioactivity of atoms. ML calculations that have been broadly utilized in drug revelation. ML algorithms and techniques are not a monolithic, homogeneous subset of AI [6, 7]. There are two primary kinds of ML calculations: Supervised and solo learning. Managed learning learns from preparing tests with realized names to decide names of new examples. Unsupervised learning perceives designs in a bunch of tests, generally without marks for the examples. The information is typically changed into a lower measurement to perceive designs in high-dimensional information utilizing unsupervised learning calculations preceding perceiving designs [7]. Measurement decrease is valuable, not just, in light of the fact that unsupervised learning is more effective in a low measurement space yet in addition on the grounds that the perceived example can be all the more effectively deciphered. Managed and unaided learning can be consolidated as semi-managed and support realizing, where the two capacities can be used for different information sets. Extensive volumes of information are basic for the turn of events, development, and practicality of effective ML calculations in each progression of the medication revelation measure. The dependence on huge superior grade information and known, very much characterized preparing sets are significantly more fundamental in exactness medication and treatments inside drug disclosure. Accuracy medication requires an extensive portrayal of all related dietary information: Genomic, transcriptomic, proteomic, and so on, to help with growing truly powerful customized medications. The far and wide utilization of high-throughput screening and sequencing, on the web multi-omic data sets, and ML calculations, in the previous twenty years, have made a prospering climate for some parts of the information age, assortment, and upkeep needed for drug improvement. The headways of information examination have effectively endeavored to depict and decipher the created information. This undertaking, upheld with ML procedures and coordinated data sets through different programming/web-application is currently routinely utilized for all means in drug revelation [1, 2, 4].

3.2.2 Dataset

We show the utilization of our AtomNet model on three sensible and testing benchmarks. Each one of these benchmarks gives an alternate and correlative appraisal of our presentation. The benefits of each are summed up here and portrayed in more detail underneath. We build our own DUD-E like benchmark to guarantee that there is no cover between the preparation and test particles. We also show a benchmark with tentatively confirmed inert particles. The DUD-E is a notable benchmark for structure-based virtual screening techniques. Momentarily, the benchmark is built by first assembling different arrangements of dynamic particles for a bunch of target proteins. The simple inclination is relieved by eliminating comparable actives; comparative actives are killed by first grouping the actives dependent on platform closeness, at that point choosing model actives from each group. At that point, every dynamic particle is combined with a bunch of property-coordinated baits (PMD). PMD is chosen to be like one another and to known activities concerning some 1-dimensional physicochemical descriptors (e.g., atomic weight) while being topologically disparate dependent on some 2D fingerprints (e.g., ECFP) [9]. The authorization of the topological difference upholds the expectation that the imitations are probably going to be dormant on the grounds that they are artificially not quite the same as any known dynamic. The benchmark comprises 102 targets, 22,886 actives (a normal of 224 actives for every objective), and 50 PMD per dynamic. We developed a DUDE-like dataset from ChEMBL 20 [17]. We considered all-action estimations that passed the accompanying channels:

- Affinity units estimated in IC50 or Ki and lower than 1 μ M.
- Target certainty was more noteworthy or equivalent to 6.
- Target has an explained restricting site in the scPDB information base and goal $\leq$ 2.5A.
- Ligands passed PAINS filers and wantonness rules.

We previously assembled target affinities by their UniProt quality name prefix and eliminated focuses for which there were under 10 dynamic ligands [10]. This sifting cycle yielded a bunch of 123,102 active and 348 targets. Second, every dynamic was matched with a bunch of 30 PMD chosen from the ZINC data set. Third, we parceled the information into preparing, approval, and testing sets by first grouping the dynamic ligands for each target in view of their Bemis-Murcko platforms and picking ligands that were at any rate 3μ M separated as the bunch models. Bunches with under 10 models were disposed of. Fourth, we characterized the test as set by arbitrarily choosing 50 focuses with their related activities and baits. Last, the preparation set was additionally parceled over the bunches into 5-crease cross-approval sets. The last dataset comprises 78,904 actives, 2,367,120 baits, and 290 targets.

Experimentally verified inactive: A restriction of benchmarks dependent on PMD is that they avoid baits that are like dynamic particles. This plan choice is to legitimize the presumption that chosen distractions are probably going to be idle, even without test approval. This implemented divergence among actives and fakes implies that PMD benchmarks come up short on some trying cases where actives and dormant particles are exceptionally comparable. We incorporate these difficult cases by subbing fakes with particles that have been tentatively approved to be idle. We developed a benchmark like our ChEMBL-20 PMD one however supplanted PMD with latent particles. We characterized an atom as latent if its deliberate movement is higher than 30μ M. We wound up with a bunch of 78,904 active, 363,187 latent, and 290 targets which were apportioned into 3-overlap get approval sets over the Bemis-Murcko groups. Targets with less than 10 clusters were never assigned into a validation set. Hence, the number of targets in the validation sets was 149.

3.2.3 Structure Based Deep Convolutional Neural Network

The network topology consists of an input layer, trailed by numerous 3D-convolutional and completely associated layers, and beat by a calculated expense layer that allocates probabilities over the dynamic and dormant classes. All units in secret layers are executed with the ReLU enactment work [1].

The info layer gets vectorized adaptations of 1A 3D frameworks set over co-edifices of the objective proteins and little particles that are tested inside the objective's limiting site [19]. To start with, we characterize the limiting site utilizing a flooding calculation cultivated by a bound ligand clarified in the scPDB information base. Second, we move the facilitation of the co-edifices to a 3D Cartesian framework beginning at the focal point-of-mass of the limiting site. Third, we test numerous presents inside the limiting site pit. Fourth, we crop the mathematical information to fit inside a suitable bounding box. In this examination, we utilized a 3D shape of 20A, focused on the source. Fifth, we interpret the input information into a fixed-size framework with 1A dividing. Every matrix cell holds a worth that addresses the presence of some essential underlying highlights in that area [5]. Fundamental primary highlights can change from basic identification of molecule types to more unpredictable protein-ligand descriptors like SPLIF, Filter, or APIF. Last, we unfurl the 3D framework into a 1D skimming point vector [20, 21]. 3D convolutional layers were carried out to help boundaries, for example, channel size. We utilized the network design of an information layer as depicted above, trailed by four convolutional layers of $128\text{x}5^3$, $256\text{x}3^3$, 256×3^3, 256×3^3 (number of channels $\times$ filer-measurement), and two completely associated layers with 1024 secret units each, bested by a strategic relapse cost layer over two movement classes.

Model Training: The model was finished utilizing stochastic inclination plummet with the AdaDelta versatile learning technique, the backpropagation calculation, and small bunches of 768 models for every slope step. No endeavor was made to improve meta-boundaries aside from the restriction of fitting the model into a GPU memory. The Preparation time was about seven days on 6 Nvidia-K10 GPUs.

Baseline method for comparison: We used Smina, a fork of AutoDock Vina, as a baseline for the SB evaluation. Smina implements an improved empirical scoring function and minimization routines over its predecessor and is freely available under the GPLv2 license.

3.3 Results and Analysis

We utilize the territory under the collector working trademark (AUC) and logAUC to report ridiculous results. The AUC shows arrangement (or positioned request) execution by estimating the territory under the bend of the genuine positive rate versus the bogus positive rate. AUC esteem of 1.0 methods wonderful detachment though an estimation of 0.5 suggests arbitrary division. LogAUC is an estimation like AUC that accentuates early enhancement execution by putting more weight toward the start of the bend so cases effectively characterized at the highest point of the position requested rundown contribute more to the score than later ones. Here, we utilized a logarithmic base of 10 which implies that the weight of the primary 1% of the positioned results is equivalent to the heaviness of the following 10%. Since the nonlinearity of a logAUC esteem makes it difficult to decipher, we deduct the region under the log-scaled irregular bend (0.14462) from a logAUC to get a changed logAUC. Thus, positive adjusted logIC esteems infer better compared to irregular execution though negative ones suggest more regrettable than irregular execution. For curtness, we will utilize changed logAUC and logAUC reciprocally for the remainder of this composition.

Table 3.1 and Figs. 3.1, 3.2, 3.3 sum up the outcomes across the three distinct benchmarks. On each of our four assessment informational indexes, AtomNet accomplishes a significant degree of improvement over Smina at a degree of exactness helpful for drug revelation. Tables 3.2 and 3.3 sum up the AUC and logAUC results regarding distinctive execution limits. On the full DUDE set, AtomNet accomplishes or then again surpasses 0.9 AUC on 59 targets (or 57.8%). Smina just accomplished

Table 3.1 Comparison of AtomNet and Smina on the DUDE, ChEMBL-20-PMD, and ChEMBL-20- inactives benchmarks. DUDE-30 refers to the held-out set of 30 targets whereas DUDE-102 refers to the full dataset.

			AUC	Adjusted	ogAUC
		Mean	Median	Mean	Median
ChEMBL-20 PMD	AtomNet	0.781	0.792	0.317	0.328
	Smina	0.552	0.544	0.04	0.021
DUDE-30	AtomNet	0.855	0.875	0.321	0.355
	Smina	0.7	0.694	0.153	0.139
DUDE-102	AtomNet	0.895	0.915	0.385	0.38
	Smina	0.696	0.707	0.138	0.132
ChEMBL-20 inactives	AtomNet	0.745	0.737	0.145	0.133
	Smina	0.607	0.607	0.054	0.044

Table 3.2 The number of targets on which AtomNet and Smina exceed given adjusted-logAUC thresholds. For example, on the CHEMBL-20 PMD set, AtomNet achieves an adjusted-logAUC of 0.3 or better for 27 targets (out of 50 possible targets). ChEMBL-20 PMD contains 50 targets, DUDE30 contains 30 targets, DUDE-102 contains 102 targets, and ChEMBL-20 inactives contain 149 targets

AUC		0.5>0.6>0.7>0.8> 0.9				
ChEMBL-20 PMD	AtomNet Smina	49 38	44 10	36 04	24 01	10 00
DUDE-30	AtomNet Smina	30 29	29 25	27 14	22 05	14 01
DUDE-102	AtomNet Smina	102 96	101 84	99 53	88 17	59 01
ChEMBL-20 inactives	AtomNet Smina	149 129	136 81	105 31	04 54	10 00

0.9 AUC for a solitary objective (wee1), around 1% of the benchmark. AtomNet accomplishes 0.8 or better AUC for 88 targets (86.3%), while Smina accomplishes it for 17 targets (16.7%). At the point when we limit the assessment to the held-out 30 objective subset of DUDE, AtomNet surpasses an AUC of 0.9 and 0.8 for 14 targets (46.7%) and 22 targets (73.3%), separately. Smina accomplishes a similar precision for 1 objective (3.3%) and 5 targets (16.7%), separately. AtomNet accomplishes mean and middle AUC of 0.855 and 0.875 on the held-out set contrasted with 0.7 and 0.694 accomplished by Smina, diminishing accessible mean blunder by 51.6%. True to form, the exhibition of AtomNet drops marginally for its held-out models, while Smina's execution doesn't.

On the ChEMBL-20-PMD dataset, AtomNet accomplishes an AUC of 0.9 or better for 10 held-out targets (20% of the set), while Smina accomplishes it on zero targets. At the point when we diminish the norm of exactness to an AUC of 0.8 or better, AtomNet prevails on 25 targets (half), while Smina just prevails on 1 target (2%). The third benchmark, which utilizes inactives rather than property-coordinated imitations, is by all accounts more testing than the other two. AtomNet predicts an AUC at or better than 0.9 for 10 targets (6.7%), while Smina prevails at nothing. For meeting or surpassing 0.8 AUC, AtomNet prevails for 45 targets (30.2%) and Smina prevails for 4 (2.7%). Albeit both Atomnet and Smina perform more regrettable than on the past benchmarks, AtomNet still essentially outflanks Smina with deference to in general and early advancement exhibitions [20]. Since this benchmark utilizes inactives it incorporates testing order instances of fundamentally comparable atoms with various marks. These cases are rejected from benchmarks utilizing PMD in light of the fact that fakes should be primarily disparate in order to assume they can be marked as latent.

Also, AtomNet shows great early advancement execution as demonstrated by the exceptionally certain logAUC esteems. AtomNet beats Smina concerning its ini-

Table 3.3 To conquer these restrictions we adopt a roundabout strategy. Rather than straightforwardly imagining channels to comprehend their specialization, we apply channels to enter information and inspect the area where they maximally fire. Utilizing this method we had the option to plan channels to substance capacities. For instance, Fig. 3.5 delineates the 3D areas at which a specific channel from our first convolutional layer fires. A visual review of the areas at which that channel is dynamic uncovers that this channel practices as a sulfonyl/sulfonamide indicator. This exhibits the capacity of the model to take in complex compound highlights from less complex ones. For this situation, the channel has deduced a significant spatial plan of info iota types with no synthetic earlier information

AUC		>0.0>0.1>0.2>0.3>				
ChEMBL-20 PMD	AtomNet Smina	49 35	44 08	36 02	27 01	10 00
DUDE-30	AtomNet Smina	30 29	27 19	22 08	17 02	10 01
DUDE-102	AtomNet Smina	102 94	99 65	88 28	69 05	43 01
ChEMBL-20 inactives	AtomNet Smina	147 123	107 35	36 05	10 00	2 00

tial enhancement, accomplishing a mean logAUC of 0.321 contrasted with 0.153 of Smina on the DUDE-30 benchmark. Imagining the ROC bends delineates the distinction between the AUC and logAUC estimations regarding the early advancement. For instance, in Fig. 3.4 we see that the AUC incentive for target 1m9m is 0.66 which may infer average execution. Be that as it may, the early enhancement shown by the logAUC for that target is 0.25 which recommends that numerous actives are amassed at the actual top of the position requested outcomes. Also, target 1qzy has AUC estimation of 0.76 however a log-scale plot recommends that 35% of its actives are gathered at the actual top of the position request list with logAUC of 0.44.

Filter visualization: Convolutional layers consist of various channels that figure out how to recognize explicit privately related highlights by consistently applying these channels across the open field. When managing pictures, one can envision these channels to confirm that the model is equipped for learning significant highlights. For instance, shown that channels in the first convolutional layer of their model could identify lines, edges, and shading inclinations. For our situation, nonetheless, we can not effectively picture the channels since: (I) the channels are 3-dimensional, and (ii) the information channels are discrete. For instance, two close RGB esteems will bring about two comparative tones however carbon is not nearer to nitrogen than to oxygen. That is, comparative qualities don't infer comparable functionalities.

ChEMBL-20 PMD BENCHMARK RESULTS

DUDE Benchmark Results

ChEMBL-20 with Inactives Benchmark Results

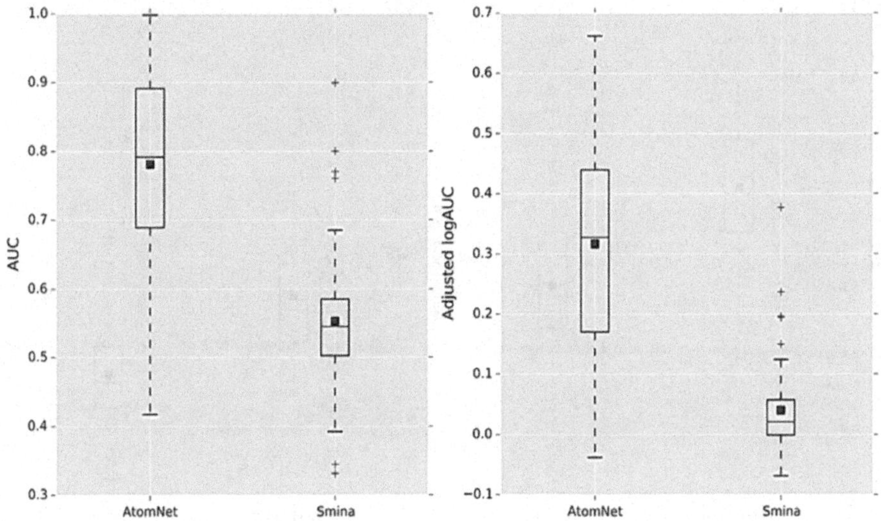

Fig. 3.1 Distribution of AUC and logAUC values of 50 ChEMBL-20-PMD targets for AtomNet and Smina

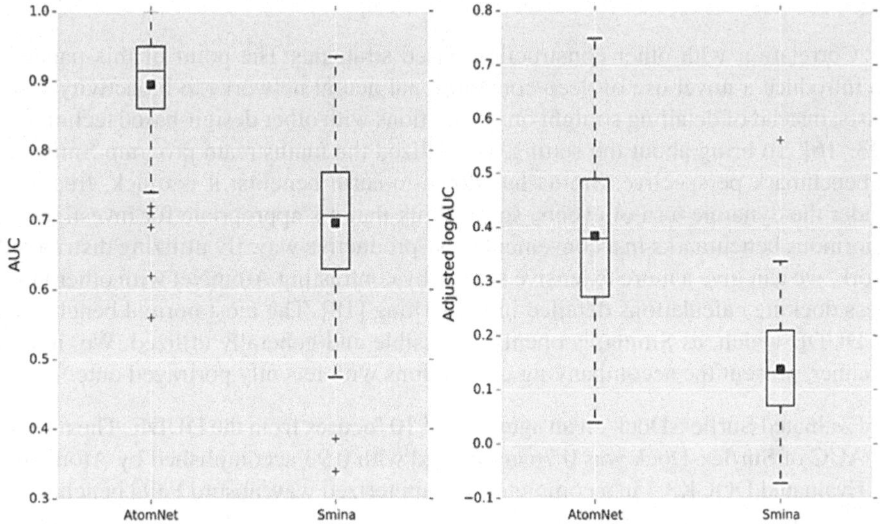

Fig. 3.2 Distribution of AUC and logAUC values of 102 DUDE targets for AtomNet and Smina

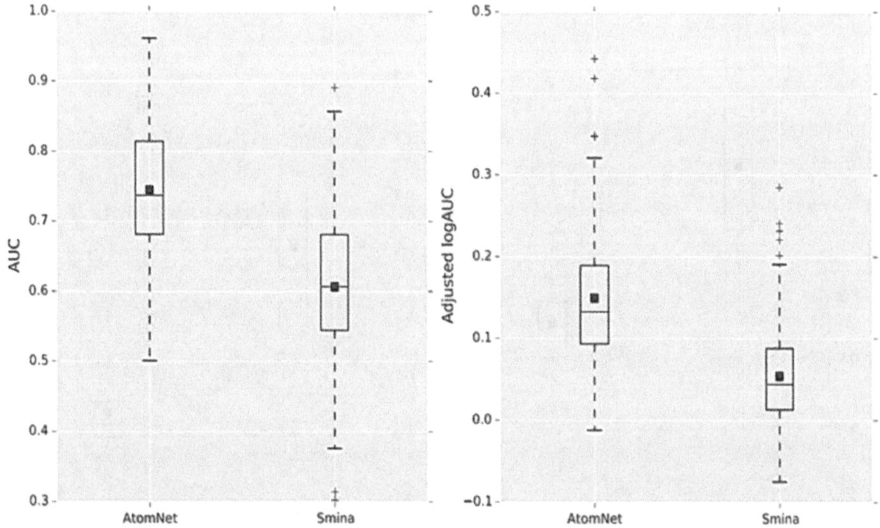

Fig. 3.3 Distribution of AUC and logAUC values of 149 ChEMBL-20-inactives targets or AtomNet and Smina

Correlation with other construction-based strategies The point of this paper is to introduce a novel use of deep convolutional neural networks to bioactivity forecasts, instead of detailing straight on correlations with other design-based techniques [13, 16]. To bring about the setting, we utilized the mainstream program Smina as a benchmark perspective. Smina has down-to-earth benefits: it is quick, free, and under the dynamic turn of events, so we think that it's appropriate for investigating enormous benchmarks in a convenient, cost-productive way. By utilizing distributed work, we can give a more extensive setting by contrasting AtomNet with other business docking calculations detailed in the writing [18]. The most normal benchmark is DUDE which, as Smina, is openly accessible and generally utilized. We, in this manner, present the accompanying correlations with recently portrayed outcomes:

- Evaluated Surflex-Dock on an agent set of 10 focuses from the DUDE. The middle AUC of Surflex-Dock was 0.76 contrasted with 0.93 accomplished by AtomNet.
- Evaluated DOCK3.7 in a completely computerized way, absurd Fella benchmark. They accomplished a mean AUC of 0.696 and logAUC of 0.174 contrasted with our AUC of 0.895 and logAUC of 0.385.
- Detailed mean AUC of 0.72 on 5 DUDE targets utilizing Dock6.7 thought to AUC of 0.852 by AtomNet.

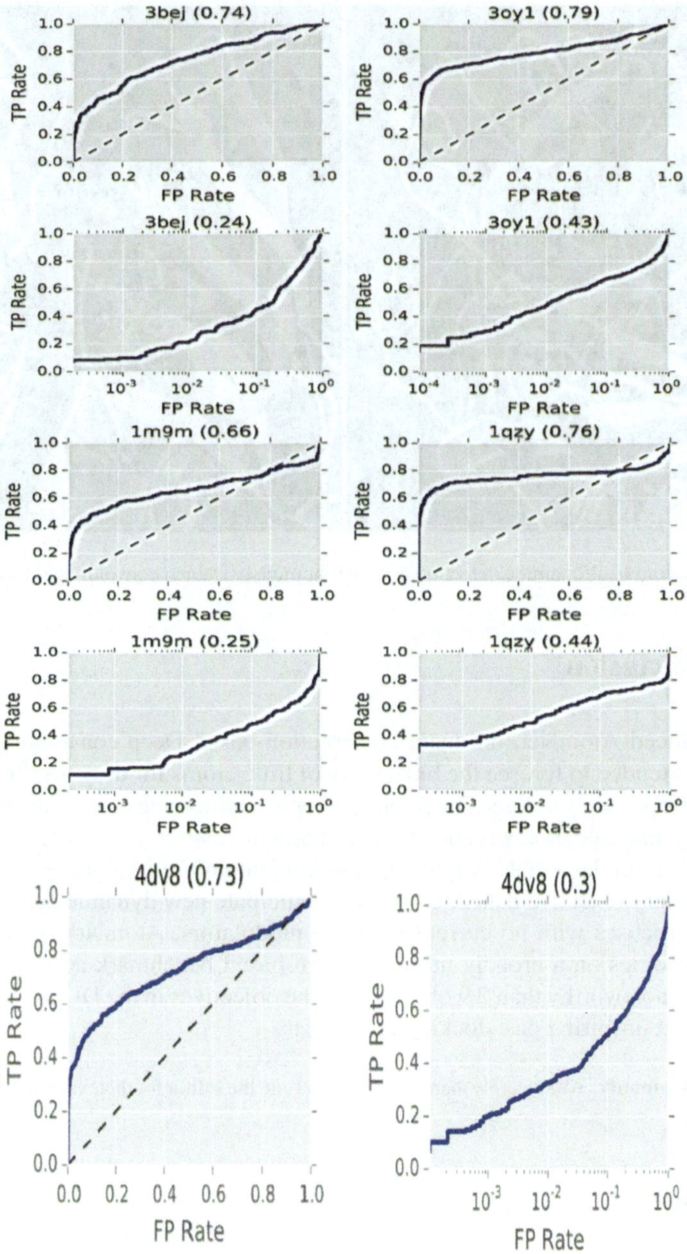

Fig. 3.4 An illustration of the differences between the AUC and logAUC measurements with respect to the early enrichment

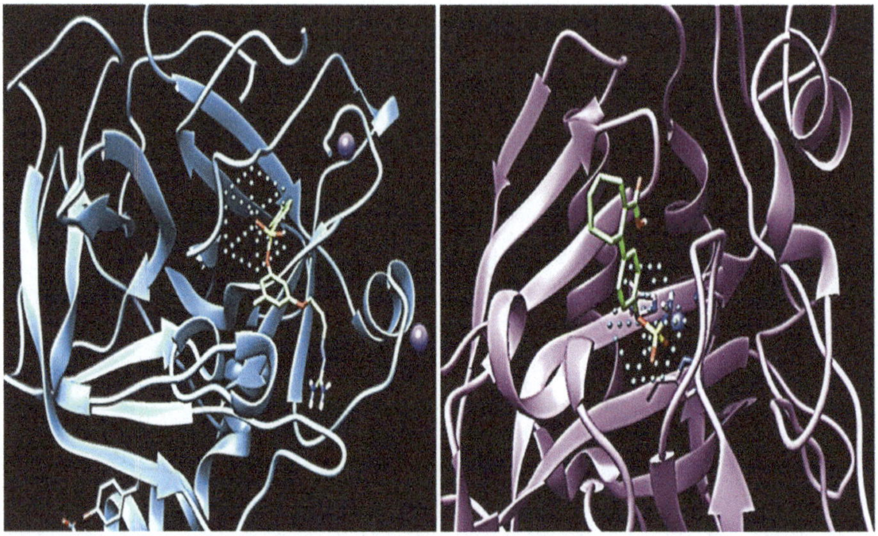

Fig. 3.5 Sulfonyl/sulfonamide detection with autonomously trained convolutional filters

3.4 Conclusion

We introduced AtomNet, the main construction-based Deep convolutional neural network, intended to foresee the bioactivity of little atoms for drug disclosure applications. The privately obliged Deep convolutional engineering permits the framework to show the intricate, non-straight wonder of atomic restriction by progressively creating general fundamental compound highlights into more mind-boggling ones. By fusing primary objective data AtomNet can anticipate new dynamic particles in any event, for focuses with no currently known modulators. AtomNet shows extraordinary outcomes on a broadly utilized design-based benchmark accomplishing an AUC more noteworthy than 0.9 on 57.8% of the objectives in the DUDE benchmark, far from awe-inspiring past docking techniques.

Acknowledgements We sincerely thank the reviewers and the Editor for their valuable suggestions.

References

1. L. Patel, T. Shukla, X. Huang, D.W. Ussery, S. Wang, Machine learning methods in drug discovery. Molecules **25**(22), 5277 (2020)
2. J. Fayyad, M.A. Jaradat, D. Gruyer, H. Najjaran, Deep learning sensor fusion for autonomous vehicle perception and localization: a review. Sensors **20**(15), 4220 (2020)
3. J. Kowalewski, A. Ray, Predicting novel drugs for SARS-CoV-2 using machine learning from a> 10 million chemical space. Heliyon **6**(8) (2020)

4. S. Mohanty, M.H.A. Rashid, M. Mridul, C. Mohanty, S. Swayamsiddha, Application of artificial intelligence in COVID-19 drug repurposing. Diabetes & Metabolic Syndrome: Clinical Research & Reviews (2020)
5. B. Ryu, D.S. Kim, A.M. DeLuca, R.M. Alani, Comprehensive expression profiling of tumor cell lines identifies molecular signatures of melanoma progression. PloS One **2**(7) (2007)
6. Y.H. Feng, S.W. Zhang, J.Y. Shi, DPDDI: a deep predictor for drug-drug interactions. BMC Bioinformatics **21**(1), 1–15 (2020)
7. F.R. Fields, S.D. Freed, K.E. Carothers, M.N. Hamid, D.E. Hammers, J.N. Ross, S.W. Lee, Novel antimicrobial peptide discovery using machine learning and biophysical selection of minimal bacteriocin domains. Drug Development Research **81**(1), 43–51 (2020)
8. C.W. Coley, R. Barzilay, T.S. Jaakkola, W.H. Green, K.F. Jensen, Prediction of organic reaction outcomes using machine learning. ACS Central Science **3**(5), 434–443 (2017)
9. Q. Yang, V. Sresht, P. Bolgar, X. Hou, J.L. Klug-McLeod, C.R. Butler, Molecular transformer unifies reaction prediction and retrosynthesis across pharma chemical space. Chem. Commun. **55**(81), 12152–12155 (2019)
10. M. Ragoza, J. Hochuli, E. Idrobo, J. Sunseri, D.R. Koes, Protein-ligand scoring with convolutional neural networks. J. Chem. Inform. Model **57**(4), 942–957 (2017)
11. P. Chang, J. Grinband, B.D. Weinberg, M. Bardis, M. Khy, G. Cadena, D. Chow, Deep-learning convolutional neural networks accurately classify genetic mutations in gliomas. Am. J. Neurorad. **39**(7), 1201–1207 (2018)
12. A. Mencattini, D. Di Giuseppe, M.C. Comes, P. Casti, F. Corsi, F.R. Bertani, E. Martinelli, Discovering the hidden messages within cell trajectories using a deep learning approach for in vitro evaluation of cancer drug treatments. Sci. Rep. **10**(1), 1–11 (2020)
13. I.I. Baskin, D. Winkler, I.V. Tetko, A renaissance of neural networks in drug discovery. Expert Opinion Drug Disc. **11**(8), 785–795 (2016)
14. F. Ghasemi, A. Mehridehnavi, A. Perez-Garrido, H. Perez-Sanchez, Neural network and deep-learning algorithms used in QSAR studies: merits and drawbacks. Drug Discov. Today **23**(10), 1784–1790 (2018)
15. H. Chen, O. Engkvist, Y. Wang, M. Olivecrona, T. Blaschke, The rise of deep learning in drug discovery. Drug Disc. Today **23**(6), 1241–1250 (2018)
16. M.H. Segler, T. Kogej, C. Tyrchan, M.P. Waller, Generating focused molecule libraries for drug discovery with recurrent neural networks. ACS Central Sci. **4**(1), 120–131 (2018)
17. H.S. Chan, H. Shan, T. Dahoun, H. Vogel, S. Yuan, Advancing drug discovery via artificial intelligence. Trends Pharmacol. Sci. **40**(8), 592–604 (2019)
18. Z. Zhou, X. Li, R.N. Zare, Optimizing chemical reactions with deep reinforcement learning. ACS Cent. Sci. **3**(12), 1337–1344 (2017)
19. Yoshimasa & Takahashi, Identification of the dual action antihypertensive drugs using tfs-based support vector machines. Chem-Bio Inform. J. **9**, 41–51 (2009)
20. G.P. Rossi, Dual ACE and NEP inhibitors: a review of the pharmacological properties of MDL 100,240. Cardiovasc. Drug Rev. **21**(1), 51–66 (2003)
21. A. Korotcov, V. Tkachenko, D.P. Russo, S. Ekins, Comparison of deep learning with multiple machine learning methods and metrics using diverse drug discovery data sets. Molecul. Pharm. **14**(12), 4462–4475 (2017)

Chapter 4
Application of AI in SCM or Supply Chain 4.0

S. P. Singh, J. Rawat, M. Mittal, I. Kumar, and C. Bhatt

Abstract Today, the enterprises that work along the supply chain management (SCM) majorly depend on their network consisting of partners and suppliers so that the moving of goods can be done effectively and efficiently. In order to increase productivity, the right technology at the right time is the only key to success. That incorporates strategies and sustainable considerations for managing and handling the various risks that one might come across during the foreground of processes. However, there is a large scope for innovation in every step involved within the supply chain by introducing artificial intelligence. These steps involve product designing from scratch, methodology of production and delivery to the required consumers. Artificial intelligence (AI) can help in optimizing and innovating these steps individually, which could exponentially improve the overall efficiency with minimum cost. This study provides the speculative allusion of application of AI in SCM, reimbursement and the challenges of AI in the SCM industry. The main endeavor of the manuscript is to explore the previously published task which might be the groundwork of upcoming research for industry.

Keywords Supply chain management · Decision making system · Inventory management · Enterprise resource planning · Artificial intelligence

4.1 Introduction

Supply chain management is a well-defined tern primarily been used for supply chain, storages, inventory management and also for logistics [1]. All businesses which are part of a supply chain are interconnected, either linked by physical or by information

S. P. Singh · I. Kumar (✉) · C. Bhatt
Graphic Era Hill University, Dehradun, Uttarakhand, India

J. Rawat
DIT University, Dehradun, Uttarakhand, India

M. Mittal
Institute of Technology & Management, Dehradun, Uttarakhand, India

S. L. Fernandes and T. K. Sharma (eds.), *Artificial Intelligence in Industrial Applications*, Learning and Analytics in Intelligent Systems 25,
https://doi.org/10.1007/978-3-030-85383-9_4

flows or by both. Transformation, warehousing of the goods and the materials, also the transportation of goods comes under the responsibilities of the physical flows which makes them the most evident, visible pieces of the supply chain. Oppositely, information flows on the other hand is responsible for the coordination of the various companies and divisions along the supply chain for their long-term plans and also to control or manage the daily flow of suppliers' materials and goods and the fluctuations in the supply chain [2, 3].

How can AI be used in the SCM?

SCM is a cumbersome issue involving a large number of micro processes which depend on each other in a complex manner. The potential for AI-driven system relies upon optimizing these steps. Machine learning (ML) as a sub-domain of AI, can introduce automation in supply chain management that can eliminate the manual and time-consuming tasks. Moreover, it can primarily bring strategies and the actions related to the business, in the play and cherish the growth as well as advancement for the betterment of the company [4–6]. Figure 4.1 shows the physical and information

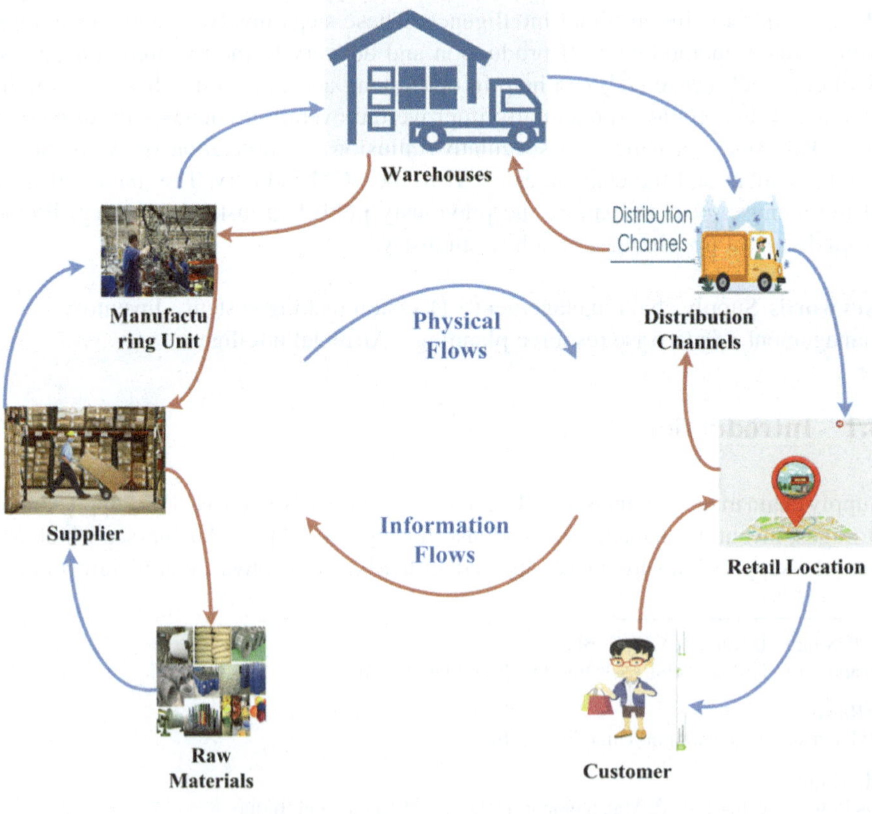

Fig. 4.1 Physical and information flow in SCM

flow in SCM and at each terminal AI can be used to improve the performance of system.

Machine Learning (ML) and Artificial Intelligence (AI) have already revolutionized the key features associated with the supply chain industry and nowadays distinguish the winners from the losers [6, 7]. These technologies help in eliminating the uncertainties as well as the unshakable inefficiencies from various phases involved in supply chain management through different methodologies and granularities that cannot be mimicked by the humans. AI plays a crucial role in supply chains like helping in order to deliver powerful and also impactful optimization capabilities which is required for planning of the capacity precisely and accurately, improving the productivity, achieving the best of the quality, lowering the costs, and overall a great and dominating output, all while encouraging to work safely and by keeping all the conditions in mind.

As per OECD (Organization for Economic Co-operation and Development), the Corona-virus pandemic might lead to the fall of global growth rate by fifty percent. This fall has already been observed throughout the wide spectra of industries and its impact on changing the working conditions of these industries cannot be overlooked [8]. World's 2nd largest economy i.e. China and some intramural supply chains have thrown as the start of corona-virus was first seen in the Wuhan city in China and thereafter dispersed in the entire world in no time, where China being responsible for the outbreak of corona-virus to other countries in the Middle East, the America, Europe, Australia and Asia, affecting all the countries and the continents mentally, socially and economically that enforced all the administration sectors of the all the countries to take preventive and sudden preventives and action to fight against the Covid-19 and also to prevent its outbreak with the country. As a result, preventive decisions and actions such as imposing restrictions on the travel either nationally or internationally, strict quarantine rules and sudden establishment of the quarantine centers on a larger scale, imposing complete lockdowns in each part of the country, etc., have led to freezing revenues, further disrupting and plummeting global food, halting critical business operations and retails and medical supply chains [9, 10].

Recently through a survey accomplished by the Institute for SCM, it came into the spotlight that nearly 75% of the companies had observed obstructions in supply-chain due to the corona-virus related restrictions, imposed by different countries. It has also affected the transportation, and is expected to see a rise in the near future [11, 37]. This is among the most serious implication arisen due to novel corona-virus pandemic and cannot be ignored.

Based on the challenges obtained from the current pandemic situations, AI can be used in SCM also. The major contributions of this work as follows:

- The extensive literature studies have been carried out on AI in Industry 4.0 and listing the research contribution from various subfield of SCM and Industry 4.0.
- This study reports the advantages of AI in SCM and reimbursement of AI-based SCM.
- Finally, this study also provides the brief description of challenges for AI in SCM.

The rest of the paper is ordered as a literature survey section as Sect. 4.2, in which major finding and limitations of past studies have been illustrated. In next Sect. 4.3, Advantages of AI in SCM are described. The reimbursement of AI-based SCM is explained in Sect. 4.4 and the brief description of challenges for AI in SCM during implementation are summarized in Sect. 4.5. Finally, Sect. 4.6 concludes the entire work.

4.2 Literature Survey

After the extensive study of literature the data examination and separation starts. The examination of literature survey is based on statistical analysis, sub-field analysis like marketing, logistics, SCM, production, etc., and AI techniques. For statistical analysis, more than 70 research articles considered between the time spam of Jan, 2008 to 20th June, 2021 for the study and considered information source was peer reviewed journal or conferences. The collected set of articles is further divided into 15 studies of marketing, 7 studies of logistic, 25 studies of production, and 27 studies of SCM. Among all studies, 26 percent taken from conference proceeding and remaining 74 percent were taken from peer reviewed journal. Figure 4.2 shows the number of articles taken from the respective years and their respective source distribution in terms of conference proceeding and peer review journal.

The brief description of previously published research article and used AI techniques is given in Table 4.1.

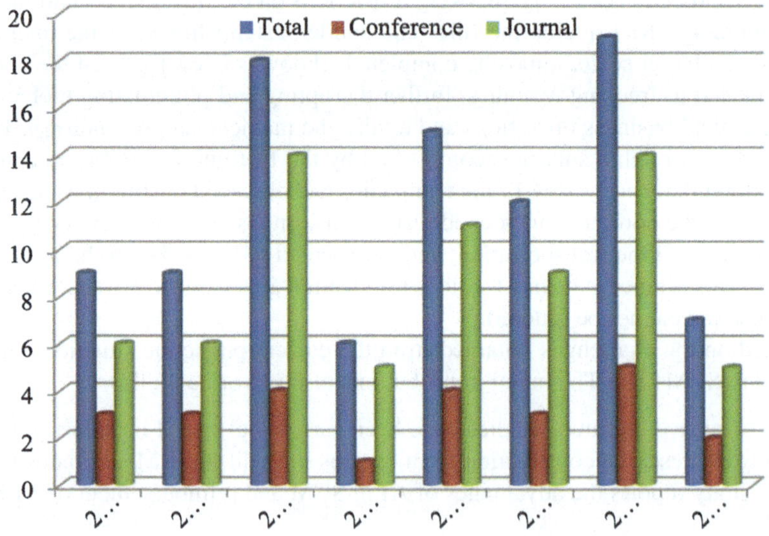

Fig. 4.2 Distribution of considered article

Table 4.1 Brief description of previously published research article and used AI techniques

Field	Subfield, [Reference no. #]	AI techniques
Marketing	Sales forecasting [12, 13, 20]	Neural network (NN) [20], Convolution neural network (CNN) [38], Genetic algorithm (GA) [39], swarm intelligence (SI), simulated annealing (SA), Decision tree (DT), Rule based system, k-mean clustering [40]
	Price forecasting [14, 21]	
	Market segmentation [15, 16, 22]	
	Customer segmentation [18, 22]	
	Decision support system [17, 18]	
	Direct marketing [17, 18, 23]	
	New product launching [19]	
Logistics	Automation system [24]	NN, SI, DT, Rule based system, k-mean clustering, Heuristic search, Robotic programming, CNN
	Terminal management system [26]	
	Inbound process management system [25]	
	Work flow management system [27]	
Production	Automation system [28]	NN, GA, SI, SA, DT, Rule based system, Expert system, CNN
	Monitoring system [29]	
	Automated decision system [30]	
	Quality monitoring system [31]	
Supply chain management	Demand and Supply forecasting [32]	NN, GA, SI, DT, Support vector machine, Data mining, CNN
	Risk management [33]	
	Network design [34]	
	Inventory management system [35]	
	Sustainability management system [36]	

From the Table 4.2, it has been observed that the four major fields in SCM industry where AI can be used are marketing, logistics, production and SCM itself. Such fields have been further divided into various subfield and huge amount of previously published articles found. For marketing, broadly used subfield is sales forecasting [12, 13, 20], price forecasting [14, 21], market segmentation [15, 16, 22], customer segmentation [18, 22], automated decision support system [17, 18], direct marketing [17, 18, 23] and new product launching [19]. For marketing field, the frequently applicable AI techniques are Neural network (NN) [20], Convolution neural network (CNN) [38], Genetic algorithm (GA) [39], swarm intelligence (SI), simulated annealing (SA), Decision tree (DT), Rule based system, k-mean clustering [40]. With the help of AI based system the performance of system is gradually increased in terms of money and various capitals with respect to time.

The application of AI based system had been also found in logistics management system. In logistic management system, AI can be used as an automation system [24], Terminal management system [26], Inbound process management system [25], Work flow management system [27] and so on. NN, SI, DT, Rule based system, k-mean clustering, Heuristic search, Robotic programming, CNN techniques are prominently found in literature for logistic management system.

For any industry, production unit is very important part and AI can be also used at production unit of any industry so that the production of the system can be increased. In AI based system the chances of error is minimal because there is no involvement of human experts. The major subfields of production where AI can be used are Automation system [28], Monitoring system [29], Automated decision system [30], Quality monitoring system [31]. In previously published research paper shown that the NN, GA, SI, SA, DT, Rule based system, Expert system, CNN techniques have been used and accounts promising results.

The application of AI is not only limited to marketing, logistics and productions. It can be also used in SCM. Various literature studies were found in SCM field with subfield like Demand and Supply forecasting [32], Risk management [33], Network design [34], Inventory management system [35], Sustainability management system [36]. The previously published article shows that the NN, GA, SI, DT, Support vector machine, Data mining, CNN AI techniques have been frequently used and yields outstanding results in the domain of SCM.

4.3 Advantages of AI iN SCM

In the recent scenario, AI plays very crucial role to make the system very much effective without the involvement of human expert. So, there are so many advantages of AI in SCM as shown in Fig. 4.3.

4.3.1 Automated Decision-Making

AI optimizes the lead SCM software and also enlarge the important aspects and the decisions by using several predictions at cognitive level and recommendations on optimal actions. This also helps in enhancement of the overall supply chain performance by accomplishing better results. Additionally, manufacturers are also helped with promising suggestions across diverse situations in form of money, time, and returns and also, by consistently learning over time, it constantly advances on these commendations as comparative circumstances change [17, 18, 31].

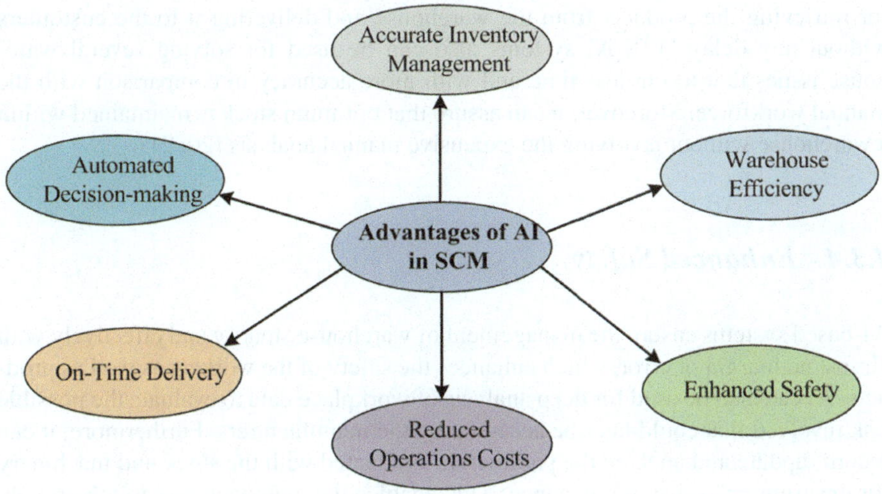

Fig. 4.3 Advantages of AI in SCM

4.3.2 Accurate Inventory Management

In order to guarantee the efficient movement of products in as well as out of the warehouse, inventory management must be accurately performed. Generally, there are numerous inventory associated variables like processing of the order, its picking and secure packing, and thus, it can turn out to be very time consuming and also resulting in high chances of the errors. Also, inventory management done accurately helps in ensuring the demand–supply equilibrium as well as avoids the problem of sudden stock-outs [19, 35].

AI-powered tools can proved to be highly effective as well as efficient in inventory management due to their ability to manage a large amount of data. These smart-tools or systems can perform the analysis and interpretation of large datasets in very less time. Moreover, they can also provide the timely guidance for predicting the demand and supply of the products through their analysis and subsequently update the system [35]. This analysis is not limited to the trends from previous year data, but also involves observing the habits of the customers to predict their future orders. Due to such intelligence of AI systems, it is now possible to avoid overstocking of inventories with unpopular products.

4.3.3 Warehouse Efficiency

Efficiency in managing the warehouse is among the most important part of supply-chain management that cannot be ignored. Automation can provide timely assistance for retrieving the products from the warehouse and delivering it to the customers without any delay [13]. AI systems also can be used for solving several warehouse issues that too in less time and with more accuracy in comparison with the manual workforce. Moreover, it can assure that optimum stock is maintained within a warehouse without involving the expensive manual analysis [20, 21].

4.3.4 *Enhanced Safety*

AI-based systems ensure the management of warehouse smartly and effectively with almost no margin of error, which enhances the safety of the workers as well as products. AI can also be used for deep analysis of workplace data to evaluate the possible risk involved that could later be accessed by the manufacturers. Furthermore, it can record, update and analyze the parameters associated with the stock and machinery for ensuring proactive maintenance. This enables the manufacturers to take timely decisions and follow safety standards for keeping the warehouse and their workers safe [24].

4.3.5 *Reduced Operations Costs*

One of the biggest and most crucial benefit of AI systems in supply-chain management is its reduced operation cost. Integrating AI systems in every phase of supply-chain management i.e. from end-to-end delivery to warehouse management, helps in cutting the cost in each of these phases [14]. These systems can work tirelessly without any break or margin-of-error which significantly reduces the overall cost and avoids workplace miss-happening. The robots performing the warehouse works are observed to be faster and accurate as compared to manual labor and has significantly achieved optimum productivity [21].

4.3.6 *On-Time Delivery*

The incorporation of AI-systems in supply-chain management has significantly optimized the time taken to deliver the product to the customers. These systems are not only capable of loading and transporting the product safely to the end-users, but are also capable analyzing the optimal route to deliver it on time. Additionally, AI-systems can analyze customer behavior to predict the upcoming orders. This has enabled the manufacturer to prepare for sudden rush in demands beforehand and store the products in warehouses to facilitate such demands. Due to this, the conventional approach of handling the demand and supply has significantly changed and has also removed the operational and manufacturing bottlenecks [23].

4.4 Reimbursement of AI-Based SCM

From the past studies, it is observed that with the help of AI and ML, an organization can witness unprecedented growth and can bring significant cost-cutting measures in its logistics. AI has optimized every phase from demand analysis to manufacturing and then from storage to end-user delivery with proper risk management and consumer grievance redressal. Due to these reasons, AI is being widely adopted by the manufacturers across all the domains of the industries for improving their performances and services to the customers.

As per McKinsey, about 61% of manufacturers have reported decrease in the costs, while about 53% have reported increase in revenues as direct result of adopting AI-systems in SCM [41]. This revenue spur has been more than 5% for about one-third of the manufacturers. Some high potential and effective areas in the management of the supply chain includes logistic optimization, spend analysis, forecasting, scheduling, planning etc.. Figure 4.4 shows the reimbursement of AI-based SCM.

4.4.1 *Bolstering Planning and Scheduling Activities*

It is often witnessed that the supply chain managers fight to set up an end-to-end progression or supply network which is more profitable and efficient than what is currently being used. This becomes more troublesome due to the factors like increasing competition, higher complexity of the network, fluctuation in demands and expanding portfolio of the products. Although, including wide range of products in one's portfolio along with critical components and spare parts, is must for smart supply-chain management. However, greater the number of products more is the need to define and characterize them, which would require various types of numerical

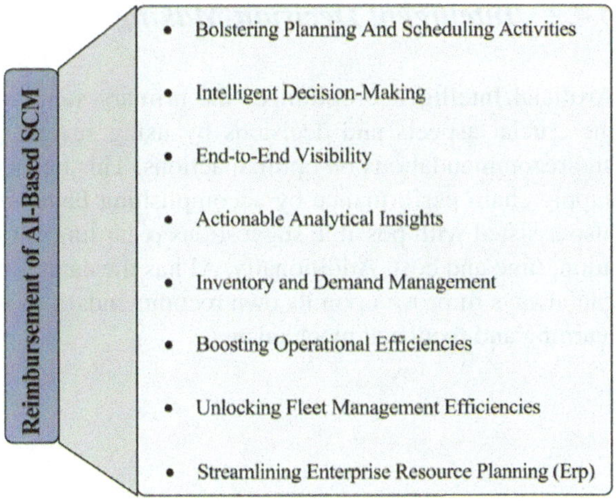

Fig. 4.4 Reimbursement of AI-based SCM

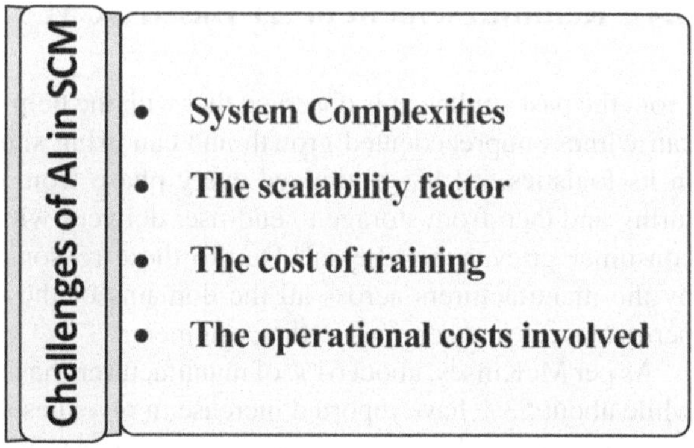

Fig. 4.5 Challenges of AI in SCM

features associated with each product. This can lead to increase in the configuration data of each of these products or components and some of these components may regularly phase-in or out. All these uncertainties or bottlenecks make it a challenging task for supply-chain manager to deduce an end-to-end process or strategy.

The implementation of AI in supply-chain management enhances the decision-making capabilities of the managers by assisting them with various tools for removing the mentioned bottlenecks. These tools can perform very complex tasks easily without consuming much time and can provide real-time analysis of current end-to-end process or supply network. The AI-systems are very efficient in accurately predicting the surge or decline in demands of certain products which helps in preparing for such scenario in advance. It can easily keep track of large number of products, or even any operational requirements or maintenance of current machinery without any errors. This makes AI among the most efficient and reliable system for integrating into the supply-chain management.

4.4.2 Intelligent Decision-Making

Artificial Intelligence optimizes the primary supply-chain software and improves the crucial aspects and decisions by using several predictions at cognitive level and recommendations on optimal actions. This helps in enhancement of the overall supply chain performance by accomplishing better results. The manufacturers are also assisted with possible suggestions regarding various aspects of revenue generation, time and cost. Additionally, AI has the capabilities of coherent learning i.e. it can always improve upon its own recommendations with the help of its continuous learning and feedback mechanism.

4.4.3 *End-To-End Visibility*

As the supply-chain network of modern globalized world is becoming complex by each passing day, it becomes important for the manufacturer to have the entire visibility and understanding of the supply-chain network. AI offers the virtualization of the huge real-time data into a single layer for the manufacturer to analyze the potential risk involved in decision-making.

4.4.4 *Actionable Analytical Insights*

Currently, there are still many organizations that lack the analytical insights for timely decision-making to generate the expected outcomes in a given time frame. The AI-driven cognitive-automation can even help in analyzing unrelated scattered data by finding the pattern between them in more efficient manner than manual analysis.

4.4.5 *Inventory and Demand Management*

Among the largest challenges that are being encountered by the organizations dealing with the supply chain is the maintaining the optimum number of stocks in their inventories. Any kind of under or over stocking would certainly lead to loss of revenues for the organization. The under stocking could let the consumer to buy the products from competitors while overstocking could cost the organization with product wastage. Moreover, it is not only the quantity of product which must be optimized but also the quantity of profit involved within an inventory. To elaborate, an organization might have multiple products in its portfolio, thus it needs to manage the inventory in terms of profit of products that are in demand. Performing such optimization manually is prone to errors and is not efficient. While AI and ML provides techniques for demand predictions by analyzing the search trends of customers belonging to a particular region. These analysis can observe the potential rise in demand of a set of products which would let the manufacturer to increase the supply of only required products and store them accordingly. The similar analysis could be done to observe the potential fall in a products' demand, which would still be helpful in avoiding any further stock piling. Similarly, these technologies can also be involved in handling the stocks and for loading or its unloading that could avoid any damage to the products during its delivery and maintains optimum level of stock in inventory.

4.4.6 Boosting Operational Efficiencies

The availability of data has never been a matter of concern for most of the organizations. However, by incorporating the IoT-based sensors into every phase of supply-chain management can provide real-time data for more accurate analysis and monitoring of these phases. A large number of these IoT sensors along with its devices are capable of sharing a large amount of data in particular interval of time, which is very difficult to analyze manually. So, AI-based techniques enable the manufacturer to evaluate the data received by the system through IoT-sensors for their meaningful interpretations. When the data is received through the supply-chain components then it becomes easy to improve the radical efficiency with the help of machine learning algorithms. ML algorithms can provide an efficient way for planning the prices of the products which depends upon the demand, product lifecycle and competition. The data received through IoT-sensors can later be gathered and reused for improving the accuracy of the algorithm.

4.4.7 Unlocking Fleet Management Efficiencies

Essentially, fleet management process is considered to be among the most neglected phase of the supply-chain management. The person responsible for managing the fleet plays an important role in establishing the supplier–consumer relation and governs the flow of products without any interruptions. Along with the shortage of labors, costs of, data overload issues, have generally been faced by the fleet managers. Managing a large fleet is an intimidating task, similar to ATC i.e. air traffic controller.

The fleet managers are responsible for finding the required information on urgent basis and then utilizing this information for decision-making. If they fail to do so then the volume of data collected in real-time would keep on increasing without being processed, adversely affecting the overall productivity. AI in logistics and supply chain provides a real-time monitoring mechanism for gaining the regular insights of the time by which the product should be delivered. It is due to the high analytical power of AI-systems over multi-dimensional data that issues of fuel inefficiency, unplanned fleet, and other bottlenecks are resolved.

4.4.8 Streamlining Enterprise Resource Planning (ERP)

In reference to a study made by panorama Consulting, about 63% of manufacturing organizations have spent an average of $3 million more than their planned ERP budgets. A supply-chain manager is involved in different business aspects like procurement, logistics and inventory management and thus cannot be helped with conventional software. Streamlining of ERP can be achieved by introducing the AI in logistics, which would help the users to connect and process the data in smart manner. It can also enhance the data processing capabilities of ERP and make it more intelligent with time to even recommend the solutions of complex problems.

4.5 Challenges of AI iN SCM

AI-systems are usually based on cloud, which makes them compatible only with high bandwidth channel and might not work efficiently in low-bandwidth environment. However, due to being the cloud-based systems, the operators do not need to install specialized hardware for accessing the functionalities of AI. Although if the operators want to install these system locally then, the initial cost might be very high.

4.5.1 System Complexities

The system complexity of the AI based system is high and it requires high bandwidth for smooth conduction of operations. Sometimes, AI based system operates on cloud based environment therefore required high configuration systems. According to the necessity of operational module, it required also specialized hardware and software system to access the AI based system. Due to this, the initial investment cost for AI based system is increased in case of SCM [42].

4.5.2 The Scalability Factor

The AI as well as cloud-based systems are highly scalable and also allow the scalability of the resource but the initial size of the system at the time of installation must be considered for producing accurate results. Since most of the AI systems are different and unique in nature, the techniques adopted by an organization must be in accordance to its requirements [42].

4.5.3 The Cost of Training

It is similar to the cost of introducing any new technology to the organization, which involves huge investment on money as well as time. However, the benefits of this training could be easily observed in later phases when high efficiency is achieved as a result of introducing it to the organization [42].

4.5.4 *The Operational Costs Involved*

The AI-based systems require high performance machines that must be regularly maintained and updated with increasing size of the data set. These machines are power consuming and a system needs to run all the time for ensuring the real-time performance, which results in high operation costs. Moreover, these systems must not be singular as, if one system fails then all its data might be lost and to avoid such situation backup is maintained, which significantly increases the overall operation cost [42].

4.6 Conclusion

Artificial intelligence provides the primary support to predictive maintenance by taking care of workplace safety and its measures to build intelligent supply chains, spot process inefficiencies that can produce the products with higher values and a better quality. AI's ability to augment, automate, enhance the customer experience, reinvent company strategies and also the decision-making, are the key aspects that makes AI the most important and holy grail of businesses being operated in SCM space. This work contributes the theoretical aspects of AI in SCM and then analyzed the previously published literature work for AI in supply chain management systems. By the consideration of literature studies, the frequently used AI techniques in supply chain management are discussed to present an inclusive perception on the existing literature in SCM. This work addressed the subfields in SCM and the suitable AI techniques for respective subfields were also discussed. Moreover, it permits the investigator to recognize what has been completed previously. This study concluded that AI in SCM can work as a booster for the exponential growth of business. Lastly, this study explains the challenges of AI in SCM implementations. These challenges can also pave the way for upcoming studies.

Funding There is no funding agencies for this research.

Declaration of Competing Interest The authors declare that they have no competing interests.

References

1. T. Davis, Effective supply chain management. Sloan Manag. Rev. **34**, 35–35 (1993)
2. M.H. Hugos, *Essentials of Supply Chain Management* (John Wiley & Sons, 2018)
3. D. Blanchard, *Supply Chain Management Best Practices* (John Wiley & Sons, 2021)
4. P. Helo, Y. Hao, Artificial intelligence in operations management and supply chain management: an exploratory case study. Product. Plann. Cont. 1–18 (2021)

5. M. Azizsafaei, D. Sarwar, L. Fassam, R. Khandan, A. Hosseinian-Far, A critical overview of food supply chain risk management. Cybersecurity, Privacy and Freedom Protection in the Connected World, 413–429 (2021)
6. D. Ni, Z. Xiao, M.K. Lim, A systematic review of the research trends of machine learning in supply chain management. Int. J. Mach. Learn. Cybern. 1–20 (2019)
7. A.P. Rathnasinghe, U. Kulatunga, H.S. Jayasena, M.K.C.S. Wijewickrama, Information flows in a BIM enabled construction project: developing an information flow model. Intell. Build. Int. 1–15 (2020)
8. M.W. Zafar, S. Saud, F. Hou, The impact of globalization and financial development on environmental quality: evidence from selected countries in the Organization for Economic Co-operation and Development (OECD). Environ. Sci. Pollut. Res. **26**(13), 13246–13262 (2019)
9. A. Kumar, W.E. Dowling, R.G. Román, A. Chaudhari, C. Gurry, T.T. Le, P.A. Kristiansen, Status report on COVID-19 vaccines development. Curr. Inf. Dis. Rep. 23(6), 1–12 (2021)
10. B. Salzberger, T. Glück, B. Ehrenstein, Successful containment of COVID-19: the WHO-Report on the COVID-19 outbreak in China (2020)
11. M. Abdirad, K. Krishnan, Industry 4.0 in logistics and supply chain management: a systematic literature review. Eng. Manag. J. 1–15 (2020)
12. W. Ketter, J. Collins, M. Gini, A. Gupta, P. Schrater, Real-time tactical and strategic sales management for intelligent agents guided by economic regimes. Inf. Syst. Res. **23**(4), 1263–1283 (2012)
13. W.I. Lee, B.Y. Shih, C.Y. Chen, Retracted: a hybrid artificial intelligence sales-forecasting system in the convenience store industry. Human Fact. Ergon. Manufac. Serv. Ind. **22**(3), 188–196 (2012)
14. T. O'Donnell, P. Humphreys, R. McIvor, L. Maguire, Reducing the negative effects of sales promotions in supply chains using genetic algorithms. Expert Syst. Appl. **36**(4), 7827–7837 (2009)
15. P.A. Sarvari, A. Ustundag, H. Takci, Performance evaluation of different customer segmentation approaches based on RFM and demographics analysis. Kybernetes (2016)
16. S. Shakya, C.M. Chin, G. Owusu, An AI-based system for pricing diverse products and services. In Research and development in intelligent systems XXVI (pp. 393–406). Springer, London (2010)
17. G. Stalidis, D. Karapistolis, A. Vafeiadis, Marketing decision support using artificial intelligence and knowledge modeling: application to tourist destination management. Procedia Soc. Behav. Sci. **175**, 106–113 (2015)
18. A.G. Rekha, M.S. Abdulla, S. Asharaf, Artificial intelligence marketing: an application of a novel lightly trained support vector data description. J. Inf. Optim. Sci. **37**(5), 681–691 (2016)
19. C.K. Kwong, H. Jiang, X.G. Luo, AI-based methodology of integrating affective design, engineering, and marketing for defining design specifications of new products. Eng. Appl. Artif. Intell. **47**, 49–60 (2016)
20. H. Pan, H. Zhou, Study on convolutional neural network and its application in data mining and sales forecasting for E-commerce. Electron. Commer. Res. **20**(2), 297–320 (2020)
21. K. Zheng, Y. Wang, K. Liu, Q. Chen, Locational marginal price forecasting: a componential and ensemble approach. IEEE Trans. Smart Grid **11**(5), 4555–4564 (2020)
22. H. Salehi, A.A. Taleizadeh, R. Tavakkoli-Moghaddam, A. Hafezalkotob, Pricing and market segmentation in an uncertain supply chain. Sādhanā **45**, 1–17 (2020)
23. M. Sarkar, A. De Bruyn, LSTM response models for direct marketing analytics: replacing feature engineering with deep learning. J. Interact. Mark. **53**, 80–95 (2021)
24. M. Klumpp, Automation and artificial intelligence in business logistics systems: human reactions and collaboration requirements. Int. J. Log. Res. Appl. **21**(3), 224–242 (2018)
25. D. Knoll, M. Prüglmeier, G. Reinhart, Predicting future inbound logistics processes using machine learning. Proc. CIRP **52**, 145–150 (2016)
26. M.A. Salido, M. Rodriguez-Molins, F. Barber, A decision support system for managing combinatorial problems in container terminals. Knowl.-Based Syst. **29**, 63–74 (2012)

27. C.K. Lee, W. Ho, G.T. Ho, H.C. Lau, Design and development of logistics workflow systems for demand management with RFID. Expert Syst. Appl. **38**(5), 5428–5437 (2011)
28. D. Sanders, A. Gegov, AI tools for use in assembly automation and some examples of recent applications. Assem. Autom. (2013)
29. A. Gligor, C.D. Dumitru, H.S. Grif, Artificial intelligence solution for managing a photovoltaic energy production unit. Proc. Manuf. **22**, 626–633 (2018)
30. F.M. Kasie, G. Bright, A. Walker, Decision support systems in manufacturing: a survey and future trends. J. Modell. Manag. (2017)
31. J. Brandenburger, V. Colla, G. Nastasi, F. Ferro, C. Schirm, J. Melcher, Big data solution for quality monitoring and improvement on flat steel production. IFAC-PapersOnLine **49**(20), 55–60 (2016)
32. T. Küfner, T.H.J. Uhlemann, B. Ziegler, Lean data in manufacturing systems: using artificial intelligence for decentralized data reduction and information extraction. Procedia CIRP **72**, 219–224 (2018)
33. Y.P. Tsang, K.L. Choy, C.H. Wu, G.T.S. Ho, C.H.Y. Lam, P.S. Koo, An internet of things (IoT)-based risk monitoring system for managing cold supply chain risks. Ind. Manag. Data Syst. **118**, 1432–1462 (2018)
34. X. Zhang, F.T. Chan, A. Adamatzky, S. Mahadevan, H. Yang, Z. Zhang, Y. Deng, An intelligent physarum solver for supply chain network design under profit maximization and oligopolistic competition. Int. J. Prod. Res. **55**(1), 244–263 (2017)
35. A.K. Sinha, W.J. Zhang, M.K. Tiwari, Co-evolutionary immuno-particle swarm optimization with penetrated hyper-mutation for distributed inventory replenishment. Eng. Appl. Artif. Intell. **25**(8), 1628–1643 (2012)
36. M. Merlino, I. Sproģe, The augmented supply Chain. Procedia Eng. **178**, 308–318 (2017)
37. H. Gupta, S. Kumar, D. Yadav, O.P. Verma, T.K. Sharma, C.W. Ahn, J.H. Lee, Data analytics and mathematical modeling for simulating the dynamics of COVID-19 epidemic—a case study of India. Electronics **10**(2), 127 (2021)
38. C. Bhatt, I. Kumar, V. Vijayakumar, K.U. Singh, A. Kumar, The state of the art of deep learning models in medical science and their challenges. Mult. Syst. 1–15 (2020)
39. P. Gupta, T.K. Sharma, D. Mehrotra, Implementation of genetic algorithm for developing knowledge centric environment in higher education. Int. J. Hybrid Intell. Syst. **14**(1–2), 13–19 (2017)
40. J. Rawat, A. Singh, H.S. Bhadauria, I. Kumar, Comparative analysis of segmentation algorithms for leukocyte extraction in the acute Lymphoblastic Leukemia images. In 2014 International Conference on Parallel, Distributed and Grid Computing (pp. 245–250). IEEE (2014, December)
41. A. Cam, M. Chui, B. Hall, Global AI Survey: AI proves its worth, but few scale impact. McKinsey Company website (2019). https://www.mckinsey.com/featured-insights/artificial-intelligence/global-ai-survey-ai-proves-its-worth-but-few-scale-impact. Accessed March 2020
42. R. Toorajipour, V. Sohrabpour, A. Nazarpour, P. Oghazi, M. Fischl, Artificial intelligence in supply chain management: a systematic literature review. J. Bus. Res. **122**, 502–517 (2021)

Chapter 5
Upskilling and Curating the Potentials of IoT Enabled Smart Cities: Use Cases and Implementation Strategies

R. Sivarethinamohan and S. Sujatha

Abstract The reason for opting the Internet of Things in the smart city system was to manage assets, resources and services efficiently and improve the operations across the city. This will be achieved by the inception of cognitive technologies, including the Internet of things (IOT). The IoT enabled Smart cities enables the utilities for improving transportation and accessibility, improve social services, promote sustainability, and give its citizens a voice. However, the barriers to smart cities are siloed, piecemeal implementations, growing expectations, uninformed citizens and shrinking budgets and little or no investment capital. Smart streetlights, lighting adapts to the activity on the street. Parking sensors provide real time information on an app where to find vacant parking spaces. Garbage sensors and automated waste collection are recent eras of smart cities. Overall it improves the operational efficiency and provides better quality of service. Thus, India thought of IoT enabled Smart cities which include housing, schools, offices and retail. In this paper, we examine significant aspects of an IoT infrastructure for smart cities, outlining the innovations implemented in the cities of India as use cases and Implementation Strategies. Exceptional attention is devoted to the potential applications of smart cities.

Keywords Service quality · Smart lighting · Smart street · Connected charging stations · Smart building · Detection sensors

R. Sivarethinamohan (✉)
Department of Professional Studies, CHRIST (Deemed to be University), Bengaluru, India

S. Sujatha
K. Ramakrishnan College of Technology, Trichy, Tamilnadu, India

S. L. Fernandes and T. K. Sharma (eds.), *Artificial Intelligence in Industrial Applications*, Learning and Analytics in Intelligent Systems 25,
https://doi.org/10.1007/978-3-030-85383-9_5

5.1 Introduction

Smart Cities are paving the way to a smart future and exponential growth of urbanization and population [1]. The first smart city in the world was arguably Amsterdam with the creation of a virtual digital city in 1994. But in India Bhubaneswar in Odisha is the 1st smart city followed by Pune-Maharashtra, Jaipur-Rajasthan, and Surat-Gujarat. Smart cities use IoT devices for example connected sensors, lights, and meters to accumulate and analyse data. IoT devices use edge computing to ensure the most relevant and important data and deliver it over the communication network. Security system of IoT is implemented to protect, monitor and control the transmission of data from the smart city network and prevent unau-thorised access to the IoT network of the city's data platform. For example, smart park-ing makes comfort drivers find a parking space and also allows for digital payment. Smart traffic management can monitor traffic flows and optimise traffic lights to reduce congestion.

A Smart city infrastructure can manage ride-sharing services energy conservation and environmental efficiencies for example streetlights that dim when the roads are empty are part of smart city initiatives. Smart city initiatives combat climate change and air pollution as well as waste management and sanita-tion via internet-enabled rubbish collection, bins and fleet management systems. It provides protection from forgery attacks [2]. Sensors monitoring high crime areas can enable an early warning for incidents like floods, landslides, hurricanes or droughts. In fact, real-time space management or structural health monitoring and feedback are built in Smart buildings to assess repairs. IBM support smart city projects by working closely with industrial partners and service operators, in France and abroad, and to provide its expertise in data collection, anal-ysis and intelligence. (i) IoT Wireless Weather Station using Arduino, ESP8266 and ThingSpeak, (ii) IoT Based Temperature and Humidity Monitoring over Thing Speak using Arduino UNO and ESP8266 and (iii) IoT Based Raspberry Pi Weather Station using DHT11 and BMP180 Sensor are some Iot projects. ThingSpeak is an open wellspring of web of things (IoT) utility and API to purchase and recover records from issues abusing the hypertext move convention and MQTT convention over the internet or through a near to space organization. ThingSpeak licenses the presentation of detecting component work programs, area following bundles, and an informal community of things with standing updates. The uprising of IoT has been recognized around the world for making life easier [3].

5.1.1 *IoT Devices*

IoT is now becoming the building block for next generation smart cities [4]. The following IoT Devices are used in the establishment of smart cities:

- Home Security.
- Activity Trackers.
- Digital Twins.

- Self-Healing Machines.
- AR Glasses.
- Ingestible Sensors.
- Smart Farming.
- Smart Contact Lense

Further Smart cities employ a variety of software, user interfaces and communication networks together with the Internet of Things (IoT).

5.1.2 How Smart Cities Work

Smart cities follow four steps:

- Collection—Smart sensors gather real-time data.
- Analysis—The data is analysed to gain insights into the operation of city services and operations.
- Communication—The results of the data analysis are communicated to de-cision makers.
- Action—Action is taken to improve operations, manage assets and improve the quality of city life for the residents.

5.1.3 Features of Smart Cities

(i) adequate water supply,
(ii) assured electricity supply,
(iii) sanitation, including solid waste management,
(iv) efficient urban mobility and public transport,
(v) affordable housing, especially for the poor,
(vi) robust IT connectivity and digitalization,
(vii) good governance, especially e-Governance and citizen participation.

In reality, IoT enabled smart city applications are applied to industry, security, retail, society, healthcare, home, energy and mobility (Fig. 5.1).

The aim of this paper is twofold: (i) Literature on smart cities was reviewed based on different applications, and (ii) we explore the main challenges and propose solutions related to the improvement of smart cities. The rest of the paper is organized as follows. In Sect. 5.2, we provide various components of the smart city and their impact in the IoT era. Section 5.3 presents challenges and opportunities of IoT enabled smart cities. In Sect. 5.4, we deepen Future of IoT enabled smart cities. Finally, in Sect. 5.5 we draw our conclusions.

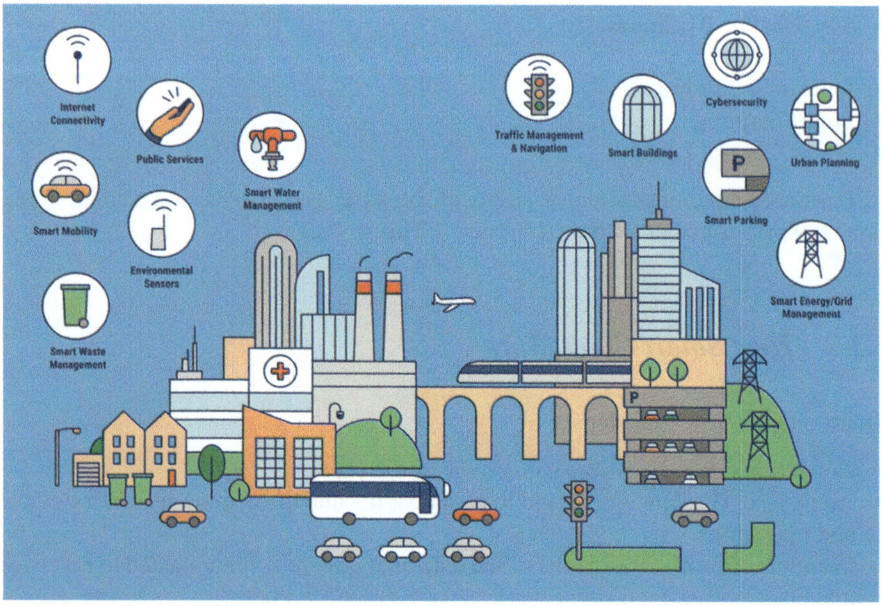

Fig. 5.1 Illustration of IoT enabled smart cities

5.2 The Various Components of the Smart City and Their Impact in the IoT Era

In fact, smart city, smart hospitals, smart highways and smart factories are interconnected with each other cloud and services and platforms like integration operation centre, smart grid, hospital optimisation, comms network optimisation, home energy management, traffic flow optimization, factory optimization and logistic optimization in addition to connected traffic cameras, automated car system, intelligent digital signage, intelligent medical devices and connected ambulances [5] (Fig. 5.2).

Potential applications of IoT technology in Smart Cities upskilled in this chapter are smart infrastructure, air quality management, traffic management, smart parking, and smart waste management etc.

5.2.1 Smart Infrastructure

Smart infrastructure provides maximum cost savings and efficiencies although it becomes a new attack surface for cyber threats. Smart Infrastructure connects energy systems, buildings and industries to proceed the best way we live and work. It is widely used from intelligent grid control and electrification to smart storage solutions, from building automation and control systems to switches, valves and sensors. In fact

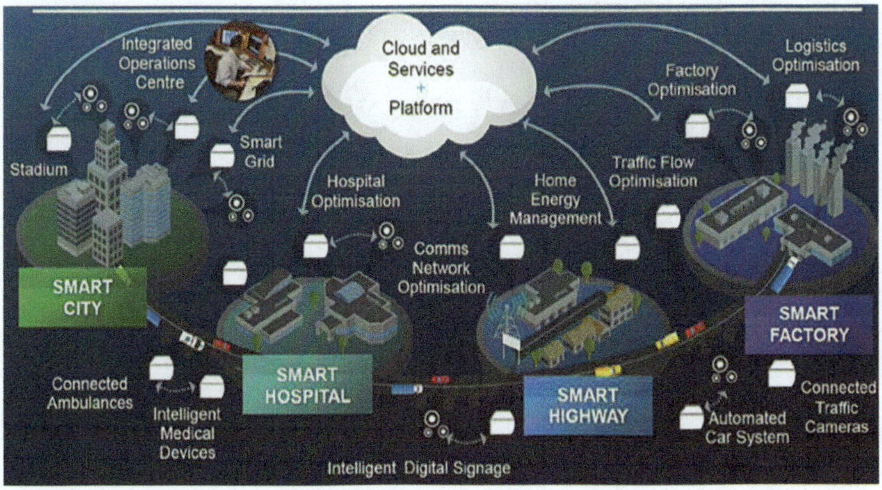

Fig. 5.2 Development of smart citie [6]

IoT enabled infrastructure are made-up of IoT platforms, access technologies, data storage and processing, data analytics, and security. Smart infrastructure in smart cities includes smart lighting, advanced connected streets, smart parking, and other transportation innovations.

5.2.1.1 Smart Lighting

Smart lighting can be used more specifically keeping real-time tracking of lighting and supplying demand-based lighting [7]. Parking lots can be dimmed during office hours, and when a car is arriving, it will be sensed and correct sectors can be illuminated, while others are kept on the diffused setting (Fig. 5.3).

5.2.1.2 Connected and Smart Streets

Connected and smart streets obtain data and supply information and services to and from many devices, which comprises details of traffic, road blockages, roadworks, etc. This ensures the well-organized administration of resources and individuals to improve public transportation and the urban landscape. A well connected smart street will have a wheelchair ramp, LED street light with panic button, walking track for the visually impaired, eco-friendly benches, bicycle track, Zebra crossing and smart signal system (Fig. 5.4).

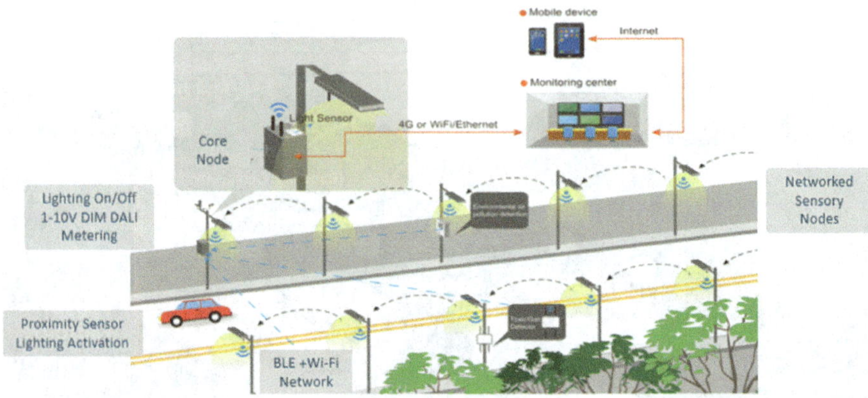

Fig. 5.3 Smart lighting architecture

Fig. 5.4 Model of smart street

5.2.1.3 Smart Parking System

Smart parking is used to locate the vacant location for a vehicle at different loca-tions. Smart parking's In-Ground Vehicle Detection Sensors provide better smart parking solutions and help drivers in the malls and city centers to explore an available parking space. Wireless sensors are embedded into parking spaces, transmitting data on the timing and duration of the space used via local signal processors into a central parking management application. Smart parking lessens congestion, drops vehicle emissions, pulls down enforcement costs, and moderates driver stress. In fact, each device is connected with the cloud servers (Fig. 5.5).

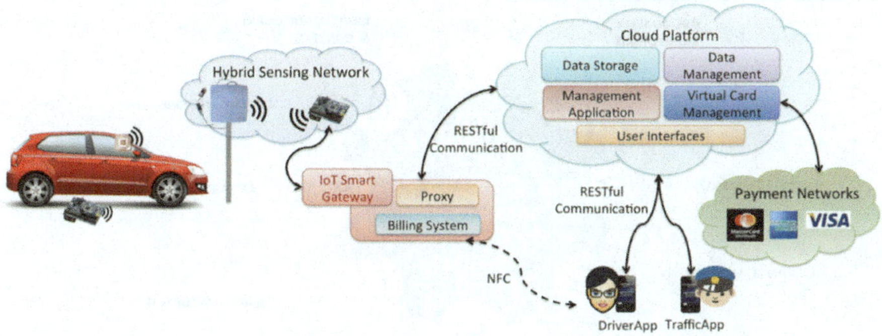

Fig. 5.5 Architecture of smart parking system

Fig. 5.6 Connected charging stations

5.2.1.4 Connected Charging Stations

Smart infrastructure manages charging stations in parking systems, city fleets, shopping malls and buildings, airports, and bus stations across the country. Electronic vehicle (EV) charging platforms are integrated with IoT to rationalize the working of EV charging and addresses the power of the power grid (Fig. 5.6).

5.2.1.5 Smart Buildings and Properties

Smart building technology shares information about what goes on in the building between systems to optimise the building's performance. This information can be used afterwards to automate several processes, from heating and ventilation to air conditioning and security. Therefore, smart building improves the indoor air quality

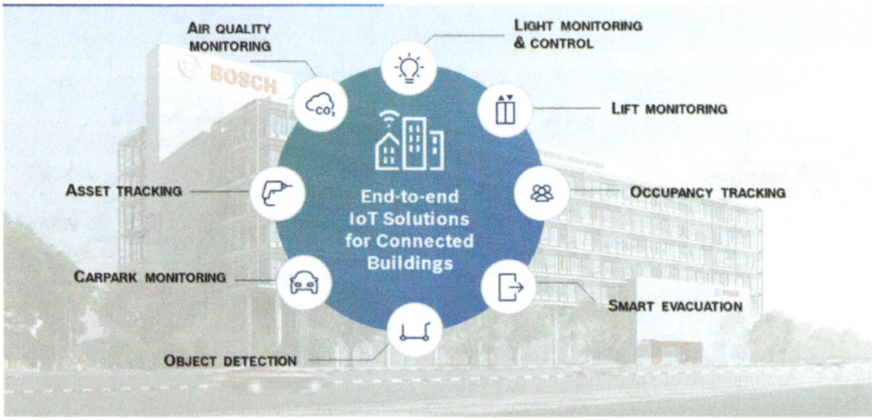

Fig. 5.7 Components of smart buildings

and lighting in a significant way. Further it enables light and lift monitoring, occupancy tracking, smart evacuation, object detection, car parking monitoring and asset tracking (Fig. 5.7).

5.2.2 Smart Industrial Environment

The industrial internet of things (IIoT) is the extension and use of the internet of things (IoT) in industrial sectors and applications. The IIoT incorporates industrial applications, containing robotics, medical devices, and software-defined production processes. It connects machines, tools, and sensors on the shop floor to ensure adequate visibility to process engineers and managers during the production. So that manufacturers can automatically track parts which move through assemblies using sensors such as RFID and break beams. Connected logistics, predictive quality and maintenance, track and trace, energy management, connected manufacturing and augmented reality apps are part of Smart Industrial Environment in the process of product innovation, supply chain, manufacturing, sales and marketing and customer services (Fig. 5.8).

5.2.3 Smart City Services

Smart cities are enabled by smart services individuals get value from consuming relevant services. City wide shared IoT sensor deployment where real-time reflexes and event responses, high value citizen centric services where near time read and respond smart services and city authorities' analytics and inputs to smart services

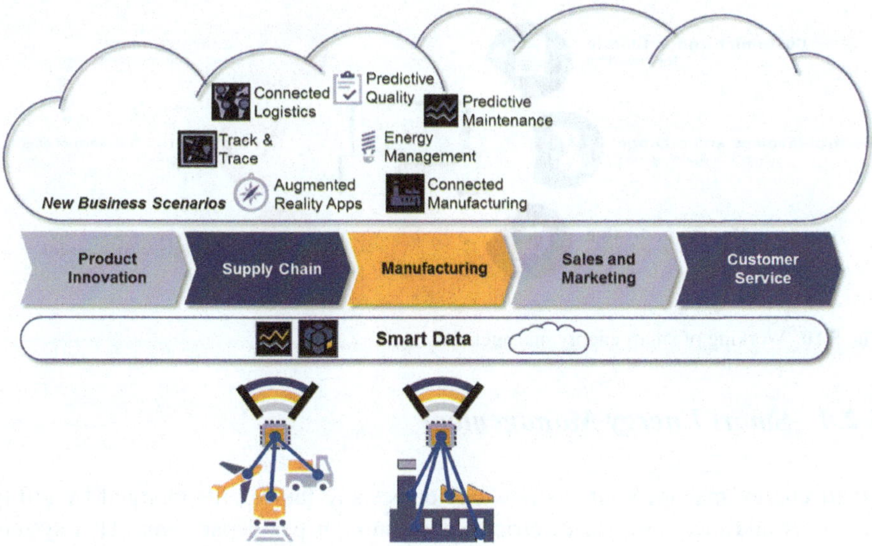

Fig. 5.8 Patern of industrial internet of things (IIoT)

such as transaction time historic trend analysis are part of smart city services which are supported by IoT gateway, data flow switch and digital asset engine [8] (Fig. 5.9).

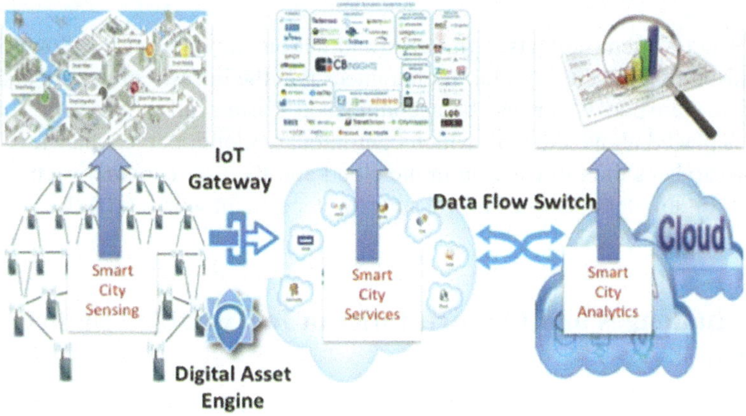

Fig. 5.9 Smart city services

Fig. 5.10 Working of smart energy management

5.2.4 Smart Energy Management

Smart energy management systems can detect any faulty bills charged by utility providers and aware of the electricity consumption per department. This system automatically analyse the power quality events easily. It is advantangious for environmental friendly and human health [9]. Figure 5.10 list out few applications and its benefits.

5.2.5 Smart Water System

Smart water system collects significant and actionable data on the flow, pressure and distribution of a city's water. It guarantees that the infrastructure and energy are utilized to transport water and be accomplished successfully. In reality The Laser sensor is placed above the tank which continuously monitors the water level in real time and updates the information in the cloud and the user can analyse the amount of water. Figure demonstrates how a smart water system works [10] (Fig. 5.11).

5.3 Challenges and Opportunities of IoT Enabled Smart Cities

Major challenging issues in the IoT is the security and privacy due to several threats, cyber attacks, risks and vulnerabilities [11]. Smart cities face challenges and opportunities are as follows.

(i) Technology challenges with coverage and capacity.
(ii) Digital security.
(iii) Legislation and policies.

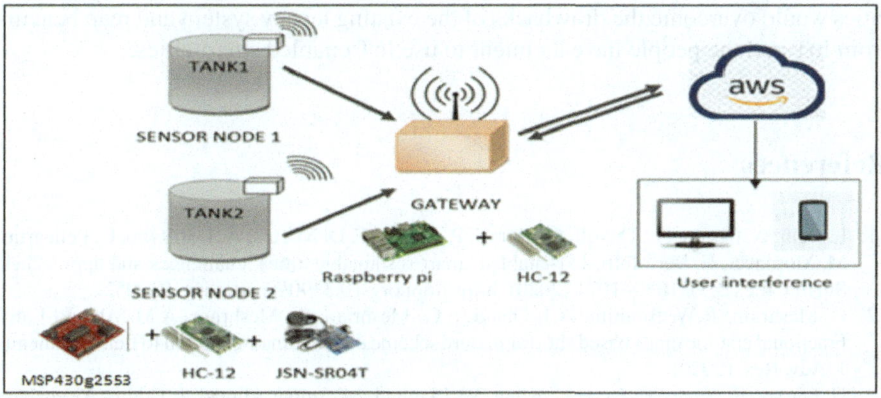

Fig. 5.11 Flow chart of smart water system

(iv) Lack of confidence or reluctance shown by citizens (lack of clarity around benefits).
(v) Funding and business models.
(vi) Interoperability.
(vii) Existing infrastructure for energy, water and transportation systems.

5.4 Future of IoT Enabled Smart Cities

The trending of "Smart Cities" was estimated to provide data-centric solutions to urban challenges. In India Smart Cities have the capacity to breed $20 trillion in economic benefits by 2026. India has a $4.5 trillion infrastructure investment by 2040 to leverage their capital, expertise, human resource, and technology. In the new ranking system, Bhopal tops the smart city list. The main emphasis of smart cities in future may be on creating an environment with zero carbon footprint that generates resources.

5.5 Conclusion

This research is an attempt to understand various aspects which smart cities are working on and targeting with the intention of effective utilization of resources and their services, with the targeted goal to develop smart ecological cities. This study proposes different applications with conceptual details. This research is aspired to discover and evaluate the significant IOT enabled applications that have affected the intention to use IoT by different users. The outcomes of the study revealed that smart cities would overcome the drawbacks of the existing legacy system and reap benefits from IoT and the people have an intent to use IoT enabled smart cities.

References

1. L. Belli, A. Cilfone, L. Davoli, G. Ferrari, P. Adorni, F. Di Nocera, A. Dall'Olio, C. Pellegrini, M. Mordacci, E. Bertolotti, IoT-enabled smart sustainable cities: challenges and approaches. Smart Cities **3**(3), 1039–1071 (2020). https://doi.org/10.3390/smartcities3030052
2. C. Meshram, R.W. Ibrahim, A.J. Obaid, S.G. Meshram, A. Meshram, A.M. Abd El-Latif, Fractional chaotic maps based short signature scheme under human-centered IoT environments. J. Adv. Res. (2020)
3. H. Qinxia, S. Nazir, M. Li, H.U. Khan, W. Lianlian, M. Sultan Ahmad, I. Uddin, AI-enabled sensing and decision-making for IoT systems. Complexity **2021**, 1–9 (2021)
4. A.H. Alavi, P. Jiao, W.G. Buttlar, N. Lajnef, Internet of things-enabled smart cities: state-of-the-art and future trends. Meas. J. Int. Meas. Confed. **129**(July), 589–606 (2018). https://doi.org/10.1016/j.measurement.2018.07.067
5. Y. Mehmood, F. Ahmad, I. Yaqoob, A. Adnane, M. Imran, S. Guizani, Internet-of-things-based smart cities: recent advances and challenges. IEEE Commun. Mag. **55**(9), 16–24 (2017). https://doi.org/10.1109/MCOM.2017.1600514
6. B. Hammi, R. Khatoun, S. Zeadally, A. Fayad, L. Khoukhi, IoT technologies for smart cities. IET Netw. **7**(1), 1–13 (2018). https://doi.org/10.1049/iet-net.2017.0163
7. A.R. Asha Rani, P. Anjali, Smart street light system using reconfigurable FPGA tool. Int. J. Res. Eng. Sci. Manag. **7**, 118–120 (2018)
8. L. Amaghionyeodiwe, Smart cities, ICT, and small businesses in the USA, in *Examining the Socio-Technical Impact of Smart Cities* (2021), pp. 155–179
9. S. Kumar, P. Tiwari, M. Zymbler, Internet of Things is a revolutionary approach for future technology enhancement: a review. J. Big Data **6**, 111 (2019). https://doi.org/10.1186/s40537-019-0268-2
10. T. Kima, C. Ramos, S. Mohammed, *Future Generation Computer Systems*, Vol. 76 (2017), pp. 159–162. https://doi.org/10.1016/j.future.2017.03.034
11. Z.B. Babovic, V. Protic, V. Milutinovic, Web performance evaluation for internet of things applications. IEEE Access **4**, 6974–6992 (2016)

Chapter 6
Identifying Impact of Sanitation and Environmental Safety in Hospitality Sector to Identify the Spread of COVID-19 Using Polynomial Function Simulation

Pushan Kumar Datta and Susanta Mitra

Abstract In the current scenario, where the world is plagued by the deadly Corona Virus Disease (Covid-19) with the virus spreading rampantly, how fast the infection can spread is a gripping question on everyone's mind, and something most of us have a good intuition for. In this paper we aim to simulate a mathematical model for predicting the rate of spread of the infectious Covid-19 disease in the dine out places and hotel occupancies where people encounter strangers in proximity. This epidemic is also expected to have a substantial influence on hotel marketing and management experts' study objectives. With the hospitality industry facing existing issues in the COVID-10 age, researchers are required to redirect their research attention to developing solutions for the industry. In our study, we will mainly focus on solving the problem on the degree of growth of the number of infected people with respect to time (in days) using simple polynomial function where we will be bringing out the function up to 5th degree. This says that the number of infected people (**N**) depends on the starting number ($\mathbf{N_0}$) and e (the natural number) raised to the product of **a** and **t**. This is called exponential growth—the driving variable, time. We have validated an expression of continuity of the viral growth using polynomial regression analysis wherein a future algorithm can be developed based on the interdependency between number of infected people and number of days.

P. K. Datta
Amity University Kolkata, Major Arterial Road (South-East), Action Area II, Newtown, Kolkata, West Bengal 700135, India
e-mail: pkdutta@kol.amity.edu

S. Mitra (✉)
The Neotia University, Diamond Harbour Road, Sarisha, West Bengal 743368, India
e-mail: susanta.mitra@tnu.in

S. L. Fernandes and T. K. Sharma (eds.), *Artificial Intelligence in Industrial Applications*, Learning and Analytics in Intelligent Systems 25,
https://doi.org/10.1007/978-3-030-85383-9_6

6.1 Introduction

COVID-19 (Coronavirus Disease 2019) pandemic caused by severe acute respiratory syndrome coronavirus 2 (SARS-CoV-2) has caused global havoc and hysteria worldwide, contributing to an international crisis [1–6]. The total impact of the COVID-19 pandemic on mortality should be the least controversial outcome to measure. However, its analysis is complicated by the lack of real time cause specific data; a potential additional issue concerns the quality of coding on the death certificates, particularly in the earlier stages of the pandemic. The disease emerged in December 2019 in Wuhan, China and spread to most continents by the beginning of March 2020 [7–9], which led to the World Health Organization declaring it as a pandemic on the 11th of March 2020. Since it was a novel disease, there was limited information on the virus. Severe Acute Respiratory Syndrome Coronavirus 2 (SARS-CoV-2) belongs to the same family as the Severe Acute Respiratory Syndrome Coronavirus 1 (SARS-CoV-1) and the Middle East Respiratory Syndrome Coronavirus (MERS-CoV). Researchers all over the world started working simultaneously to understand the virus to provide the necessary treatment regime or vaccine to reduce its impact on victim [8]. In this context, estimating excess deaths for all causes with respect to past year trends has been used in several countries as an effective way to evaluate the total burden of the COVID-19 pandemic, including direct COVID-19-related as well as indirect effects. This approach can only present global pictures of the total burden of the first wave of the infection [10]. Nevertheless, to understand the dynamics of the pandemic, we need to analyze data at sub-national level; this would allow to account for geographical differences due to the infectious nature of the disease, as well as those in the population characteristics and health system provision [11–15]. Recently, health authorities of each country have been urged to release mortality data at high spatial and temporal resolution to evaluate the pattern of the disease and to compare it with the expected trend from previous years.

6.2 Meaning of the Study: Possible Explanations and Implications for Clinicians and Policymakers

The main objective of this study is to discuss COVID-19 and its possible implications on food, water, sanitation, hygiene and environmental safety during the pandemic and the impact it will have now and future. In 2014, India's government started the "Swachh Bharat Mission," or "Clean India Mission," which is the country's largest sanitation and cleaning initiative. The government declared India open defecation free (ODF) in October 2019, yet a recent study by Gupta et al. (2020) using data from the rural sanitation survey 2018 (collected between August and December 2018) of four states found that the country is still far from being ODF (Utter Pradesh, Bihar, and Rajasthan) [16–19]. One of the main issues that hospitality industry needs to answer is how customers feel about visiting a restaurant or hotel during a coronavirus epidemic. Is it safe to say they are ready to return? If not, what will entice them back? It's worth noting that RT-qPCR analysis is heavily used in numerous investigations of the SARS-CoV-2 virus. On January 30, 2020, the first case of COVID-19 in India, which originated in China, was recorded. Chemical dangers include the presence of chemical agents such as formalin, DDT, disinfectants, and other disinfectants, as well as a greater proportion of recommended chemicals than the daily dose. Biological risks refer to the presence of microorganisms (bacteria, viruses, fungus, and protozoa) in food, as well as their toxins and other biological agents that may be harmful to humans [20–22]. With over 8 million confirmed instances of COVID-19 infection, India presently has the largest number of confirmed cases in Asia and the second-highest number of confirmed cases in the world after the United States. COVID-19 is a highly contagious disease that results in a very rapid transition from one person to another and has led to death in the most susceptible, especially those older than 60 years and people with underlying conditions. In the COVID-19 business climate, hospitality firms are required to make significant adjustments to their operations to protect staff and customer health and safety, as well as increase consumer willingness to patronise their business. These data show that, on general, people are still hesitant to eat at a sit-down restaurant, travel to a destination, and stay in a hotel. Because the hotel industry's breakeven point is extremely high due to high operational expenses, many hospitality firms rely significantly on increased demand for their services and goods to survive. In the first wave, this technique can only provide a global picture of the total infectious burden. However, data at the subnational level must be analysed to properly grasp the pandemic's dynamics; this will allow us to account for geographical variations due to the disease's infectious nature, as well as inequalities in demographic features and health system provision. Artificial Intelligence and data science has paved the way for myriad possibilities in terms of technology enhancement. Today, most data analysers and scientists use machine learning, which is a branch of AI, to analyse data in diverse domains. Data mining researchers and data scientists have contributed majorly understanding the behaviour of COVID-2019 and for acquiring technologies and related data on the nature of this virus. According to data analysis, the propagation of each epidemic is exponential up to a point, beyond which the infection rate remains relatively constant. Since a result, flattening the top of the curve is critical for efficient epidemic management, as it allows hospitals to deal with existing resources until the pandemic is finished. In this proposed study, we aim to predict the future spread of COVID19 infection in India using machine learning algorithm based exponential polynomial regression analysis up to 5th degree polynomial. Regression analysis is a method of predictive modelling technique that examines the relationship between a dependent and an independent variable. In statistics, polynomial regression is a type of regression analysis in which the relationship between the independent variable x and the dependent variable y is represented as the nth degree polynomial in x. This type of study will help to make the correct decision on this epidemic and to work out a clear strategy for future measures so that we will be able to better treat and improve infrastructure, well-being, vaccine production and other epidemics.

6.3 System Model and Algorithm

The field of machine learning has given us a huge space, especially in the field of data analysis. Hence, we concentrate on solving our problem in the field of data analysis which is a crucial part of machine learning. This simulation-based approach to the post-processing of the model output allows to fully characterize the uncertainty in the mortality rates and to propagate it in a principled way through to the predicted mortality counts. We present plots of observed and predicted trends in mortality rates as well as maps and plots of percentage excess mortality, estimated as the difference between the observed number of effects and the predicted number under current trends (based on past data and on the assumption that the pandemic did not occur), divided by the observed number. While the COVID-19 epidemic has presented unknown problems to the hospitality sector and academia, it has also provided excellent study possibilities for hospitality academics. In comparison to past crises, the scope of this crisis and its destructive impact on business, employees, and consumers is unprecedented.

6.4 Machine Learning Approach as a Strength and Weakness of the Study

To capture the long-term temporal trend, we considered a year-specific random intercept and accounted for the potential non-linearity in the mortality relative risks through a weekly smooth effect, where each week is assumed to depend on the value of the previous one (order one random walk) [23–25]. As the temporal trend might be different in space, we allowed a different random walk for each province. Finally, as there is extensive evidence of a non-linear effect of temperature on mortality, we included a smoothed effect on weekly temperature at municipality level, through a second order random walk, so that each week is influenced by the previous two [26–29]. The method involves a 2-step mechanism which involves training of the system and consequent testing and if required further rectification. But the main motive will anyhow be to make our system model learn automatically by systematic feeding process, so that it can perform automatically later with high accuracy. But this process follows some specific algorithms based on the type of problem we want to solve. In our problem we will use Regression technique which is one of the most powerful tools under the supervised technique of machine learning. Regression in simple words describes the linear relationship between various independent and dependent variables, which is used in prediction of some known parameters. There are various types of Regression algorithms like: Linear, Polynomial, Logistic and so on. Our aim is to predict the future rate of infected people due to COVID-19 and hence our work is to predict, so definitely we will be using regression technique. Here we are using polynomial regression which is a special type of linear regression based on some degrees of polynomial equations. This will give us a curvilinear relationship between the number of days considered and number of cases each day. For this, we took 5 degrees of polynomial equation and got the best fit line for our model [30–34]. But the model will not respond successfully with higher than 5th degree polynomial regression.

Let us first consider a linear equation.

$$y = mx + c \rightarrow$$

This is a polynomial equation of degree 1. Here, x is the independent variable and y is the dependent variable, c is the intercept on y axis while m is the slope of the line. As mentioned earlier, we have performed one degree, two degree, three degree, four degree and fifth degree polynomial, where we found that the fifth degree polynomial regression gave us the best result with higher accuracy (Figs. 6.1, 6.2, 6.3, 6.4 and 6.5).

1st degree polynomial:

$$y = mx + c$$

2nd degree polynomial:

$$y = m_1x^2 + m_2^x + c$$

Similarly,

Fifth degree polynomial expression:

$$y = m_1x^5 + m_2x^4 + m_3x^3 + m_4x^2 + m_5x^1 + c$$

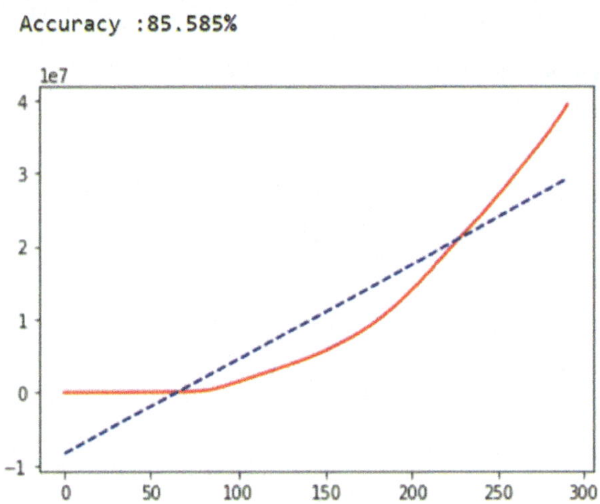

Fig. 6.1 1st degree coefficients

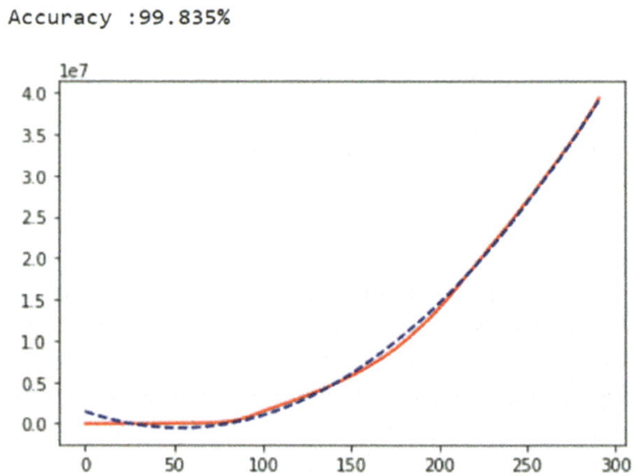

Fig. 6.2 2nd degree coefficients

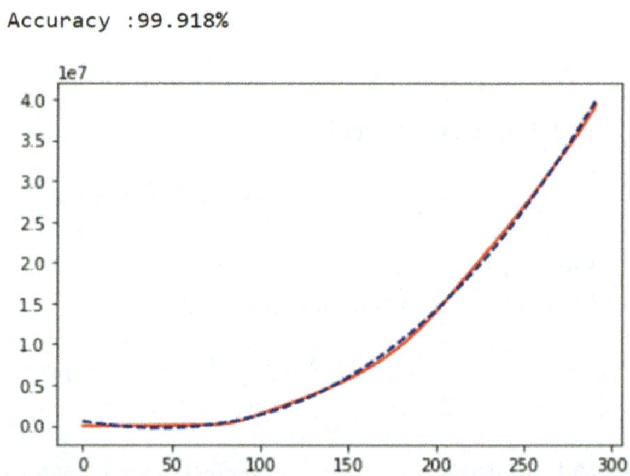

Fig. 6.3 3rd degree coefficients

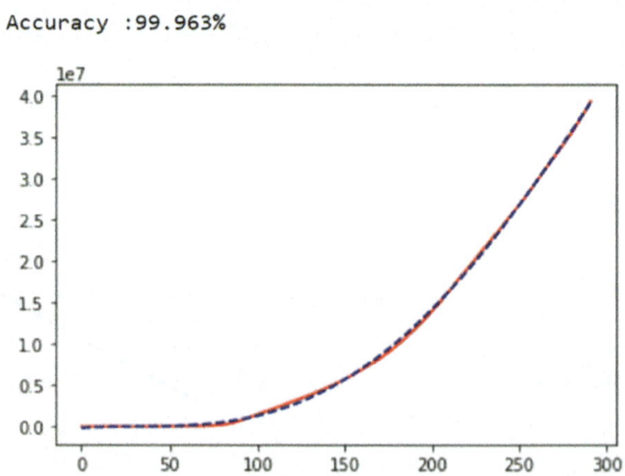

Fig. 6.4 4th degree coefficients

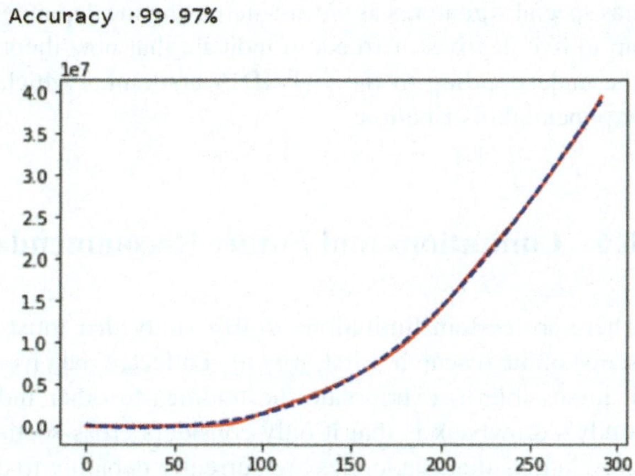

Fig. 6.5 5th degree coefficients

Hence, we applied the above-mentioned regression model for our analysis. In addition to providing estimates for the specific Indian case, the proposed modelling framework is highly scalable and adaptable. A natural extension of our work consists in the analysis of cause-specific mortality, when the data become available: this will lead to a better understanding of the trends in COVID-19-related mortality, with respect to other competing mortality risks. The above plots clearly show how our model responded to the best solution for 5th degree polynomial, showing the best fit regression line. While preliminary findings indicate that visible sanitization efforts (such as hand sanitizers at the entrance, staff wearing masks and gloves), social distancing, limiting the number of customers served, more rigorous and frequent cleaning of high-touch surfaces in common areas, and employee training of health and safety protocols are the most important, preliminary findings also indicate that implementing social distancing, limiting the number of customers served, and more rigorous and frequent cleaning of high-touch surfaces in common areas, and employee training and limiting the number of customers served, and more rigorous and frequent cleaning of high-touch surfaces in common areas.

6.5 Conclusion

We concentrate here on a very basic methodology called Polynomial Regression Analysis, in which the relationship between the independent variable x and the dependent variable y is modelled as a polynomial of the nth degree in x. A nonlinear relationship between the value of x and the corresponding conditional mean of y, denoted $E(y)$, matches the polynomial regression. Though polynomial regression suits the data with a nonlinear model, it is linear as a statistical estimation problem. The results from this simple analysis seem to show that epidemics have a similar and predictable pattern for all different countries but also revealed COVID19 epidemics has special signatures as we are increasing the dependency. Our dependency analysis up to five degrees of freedom indicate that new theoretical models are in need for the understanding of the COVID19 epidemics which are not random but have an exponential distribution.

6.6 Limitations and Future Recommendations

There are certain limitations to this study that must be addressed to broaden the scope of the research. First, we only collected data from industry personnel, making it impossible to extrapolate the findings to other industries [35–40]. Second, the study's drawback is that it only considers cross-sectional design since it examines correlations that academics are currently debating to determine their direction [41–46]. Some studies have shown that employability predicts work insecurity, whereas others have found that job instability predicts employability. Similarly, the starting state of mental health may have an influence on economic crisis worries, and COVID-19 is still up for debate. These recommendations may create new possibilities for future researchers, and they may be taken into account in the future. Third, perceived job instability is used as a mediator in this study; however, future researchers might explore using a different mediator [47–50]. The changeover in conditions at the time of writing this essay might be the fourth restriction. Because the data was obtained at a time when public anxiety of COVID-19 was at an all-time high, additional associated circumstances such as lockdown and quarantine were in effect in numerous areas. As a result, previous research suggests that a high fear level might lead to a different outcome. Fifth, COVID-19 anxiety is a modern phenomenon in the literature. As a result, future researchers should examine our model in the context of various countries to verify or reject the recent evidence. Furthermore, a comparing research to examine the effect of COVID-19 [51–55] fear across various age groups, genders, and work status is recommended. Sixth, this research focuses solely on the economic and psychological aspects that influence employees' mental health [56–59]. It does not address the organisational viewpoint during a crisis, such as the tactics companies use to deal with the issue by meeting employee demands and increasing profitability; this is an area that future scholars could investigate more. Finally, because this study is based on a self-reported questionnaire gathered at a single moment in time, the findings' generalizability may be questioned. Therefore, it is suggested that the future researcher adopt a longitudinal study strategy to confirm the consistency of the current findings.

References

1. M.D. Adams, Air pollution in Ontario, Canada during the COVID-19 state of emergency. Sci. Total Environ. **742**, 140516 (2020)

2. W. Ahmed, N. Angel, J. Edson, K. Bibby, A. Bivins, J.W O'Brier, First confirmed detection of SARS-CoV-2 in untreated wastewater in Australia: a proof of concept for the wastewater surveillance of COVID-19 in the community. Sci. Total Environ. **728**, 138764 (2020)
3. J.D. Berman, K. Edisu, Changes in U.S. air pollution during the COVID-19 pandemic. Sci. Total Environ. **739**, 139864 (2020)
4. A. Biswal, T. Singh, V. Singh, K. Ravindra, S. Mor, COVID-19 lockdown and its impact on tropospheric NO_2 concentrations over India using satellite-based data. Heliyon **6** (2020)
5. M. Bodrud-Doza, S.M.D. Islam, T. Rume, S.B. Quraishi, M.S. Rahman, M.A.H. Bhuiyan, Groundwater quality and human health risk assessment for safe and sustainable water supply of Dhaka Citydwellers in Bangladesh. Groundwater Sustain. Develop. **10**, 100374 (2020)
6. P. Caine, Environmental impact of COVID-19 lockdowns seen from space. Sci. Nat. (2020). https://news.wttw.com/2020/04/02/environmental-impact-covid-19-lockdowns-seen-space 2 April 2020
7. J. Calma, The COVID-19 pandemic is generating tons of medical waste. The Verge, Mar. 2020;26 (2020). https://www.theverge.com/2020/3/26/21194647/the-covid-19-pandemic-is-generating-tons-of-medical-waste
8. N. Cellini, N. Canale, G. Mioni, S. Costa, Changes in sleep pattern, sense of time, and digital media use during COVID-19 lockdown in Italy. PsyArXiv. (2020)
9. I. Chakraborty, P. Maity, COVID-19 outbreak: migration, effects on society, global environment and prevention. Sci. Total Environ. **728**, 138882 (2020)
10. N. Chen, Epidemiological and clinical characteristics of 99 cases of 2019 novel coronavirus pneumonia in Wuhan, China: a descriptive study. Lancet. **395**, 507–513 (2020)
11. C. Clifford, The water in Venice, Italy's canals is running clear amid the COVID-19 lockdown (2020). https://www.cnbc.com/2020/03/18/photos-water-in-venice-italys-canals-clear-amid-covid-19-lockdown.html
12. Climate and Clean Air Coalition (CCAC) United Nations environment Programme (UNEP) (2019). https://ccacoalition.org/en/resources/25-clean-air-measures-asia-and-pacific
13. CPCB Central Pollution Control Board, Ministry of Environment, Forest and Climate Change, Government of India. Daily River Water Quality Monitoring Data (2020)
14. R. Cooper, Institute of Development Studies; Brighton, UK: 2020. Water Security beyond Covid-19. K4D Helpdesk Report 803. https://opendocs.ids.ac.uk/opendocs/handle/20.500.12413/15240
15. CREA Air quality improvements due to COVID-19 lock-down in India. Centre for Research on Energy and Clean Air (2020). https://energyandcleanair.org/air-quality-improvements-due-to-covid-19-lock-down-in-india/ 16 April 2020
16. K. Cripps, Thailand's most popular island goes into lockdown as Covid-19 cases surge. CNN travel, CNN (2020). https://edition.cnn.com/travel/article/phuket-thailand-lockdown/index.html 10 April 2020
17. ECDC European Centre for Disease Prevention and Control (ECDC). COVID-19 situation update worldwide, as of 14 July (2020). https://www.ecdc.europa.eu/en/geographical-distribution-2019-ncov-cases
18. O. Ellabban, H. Abu-Rub, F. Blaabjerg, Renewable energy resources: current status, future prospects and their enabling technology. Renew. Sustain. Energy Rev. **39**, 748–764 (2014)
19. S. Evans, Global emissions analysis: coronavirus set to cause largest ever annual fall in CO_2 emissions. Carbon Brief (2020). https://www.carbonbrief.org/analysis-coronavirus-set-to-cause-largest-ever-annual-fall-in-co2-emissions 4 September 2020
20. O.O. Fadare, E.D. Okoffo, Covid-19 face masks: a potential source of microplastic fibers in the environment. Sci. Total Environ. **737**, 140279 (2020)
21. I. Ghosh, The emissions impact of coronavirus lockdowns, as shown by satellites (2020) https://www.visualcapitalist.com/coronavirus-lockdowns-emissions/
22. L. Goines, L. Hagler, Noise pollution: a modern plague. South. Med. J. **100**(3):287–294 (2007)
23. M.L. Holshue, First case of 2019 novel coronavirus in the United States. N. Engl. J. Med. **382**, 929–936 (2020)

24. C. Huang, Y. Wang, X. Li, L. Ren, J. Zhao, Y. Hu, L. Zhang, G. Fan, J. Xu, X. Gu, Clinical features of patients infected with 2019 novel coronavirus in Wuhan, China. Lancet. **395**, 497–506 (2020)
25. D.S. Hui, E. Azhar, T.A. Madani, F. Ntoumi, R. Kock, O. Dar, The continuing 2019-nCoV epidemic threat of novel coronaviruses to global health-The latest 2019 novel coronavirus outbreak in Wuhan, China. Int. J. Infect. Dis. **91**, 264–266 (2020)
26. E. Hysa, A. Kruja, N.U. Rehman, R. Laurenti, Circular economy innovation and environmental sustainability impact on economic growth: an integrated model for sustainable development. Sustainability **12**, 4831 (2020)
27. IEA, The International Energy Agency; Paris, France: 2020. Oil Market Report: March (2020). https://www.iea.org/reports/oil-market-report-march-2020
28. India Environment Portal (IEP) Impact of lockdown (25th March to 15th April) on air quality (2020). http://www.indiaenvironmentportal.org.in/content/467415/impact-of-lockdown-25th-march-to-15th-april-on-air-quality/
29. S.M.D. Islam, M.A.H. Bhuiyan, Impact scenarios of shrimp farming in coastal region of Bangladesh: an approach of an ecological model for sustainable management. Aquacult. Int. **24**(4), 1163–1190 (2016)
30. S.M.D. Islam, M.A.H. Bhuiyan, Sundarbans mangrove forest of Bangladesh: causes of degradation and sustainable management options. Environ. Sustain. **1**, 113–131 (2018)
31. S.M.D. Islam, M. Bodrud-Doza, R.M. Khan, M.A. Haque, M.A. Mamun, Exploring COVID-19 stress and its factors in Bangladesh: a perception-based study. Heliyon. **6**(7) 2020
32. S.M.D. Islam, M.E. Huda, Water pollution by industrial effluent and phytoplankton diversity of Shitalakhya River, Bangladesh. J. Sci. Res. **8**(2), 191–198 (2016)
33. S. Jribi, H.B. Ismail, D. Doggui, H. Debbabi, COVID-19 virus outbreak lockdown: what impacts on household food wastage? Environ. Dev. Sustain
34. E. Kerns, E.A. Masterson, C.L. Themann, G.M. Calvert, Cardiovascular conditions, hearing difficulty, and occupational noise exposure within US industries and occupations. Am. J. Ind. Med. **61**(6), 477–491 2018
35. M. Lenzen, Y.Y. Sun, F. Faturay, Y.P. Ting, A. Geschke, A. Malik, The carbon footprint of global tourism. Nat. Clim. Change. **8**, 522–528 (2018)
36. B. Ma, X. Li, Z. Jiang, J. Jiang, Recycle more, waste more? When recycling efforts increase resource consumption. J. Clean. Prod. **206**, 870–877 (2019)
37. Y. Ma, X. Lin, A. Wu, Q. Huang, X. Li, J. Yan, Suggested guidelines for emergency treatment of medical waste during COVID-19: Chinese experience. Waste Dispos. Sustain. Energy. **2**, 81–84 (2020)
38. S. Mallapaty, How sewage could reveal true scale of coronavirus outbreak. Nature. **580**, 176–177 (2020)
39. L.Y.K. Nakada, R.C. Urban, COVID-19 pandemic: impacts on the air quality during the partial lockdown in São Paulo state. Brazil. Sci. Tot. Environ. **730**, 139087 (2020)
40. J. Pan, *China's Environmental Governing and Ecological Civilization. China Insights* (Springer, Berlin, Heidelberg, 2016). Sustainable industrialization
41. R.P.T. Pereira, G.M. Ribeiro, V. Filimonau, The carbon footprint appraisal of local visitor travel in Brazil: a case of the Rio de Janeiro–São Paulo itinerary. J. Clean. Prod. **141**, 256–266 (2017)
42. R. Puig, E. Kiliç, A. Navarro, J. Albertí, L. Chacón, P. Fullana-i-Palmer, Inventory analysis and carbon footprint of coastland-hotel services: a Spanish case study. Sci. Total Environ. **595**, 244–254 (2017)
43. M. Rahman, Rare dolphin sighting as Cox's Bazar lockdown under COVID-19 coronavirus (2020). https://www.youtube.com/watch?v=gjw8ZlIIlbQ
44. M.M. Rahman, M. Bodrud-Doza, M.D. Griffiths, M.A. Mamun, The lancel global health. Biomedical Waste amid COVID-19: Perspectives from Bangladesh (2020)
45. S.H. Rahman, S.M.D. Islam, Degrading riverine ecology of Bangladesh and options for management. SUB J. Sustain. Environ. Develop. **1**, 11–27 (2016)
46. Saadat S., Rawtani D., Mustansar C. Hussain environmental perspective of COVID-19. Sci. Total Environ. **728**:138870 (2020)

47. A.N. Sajed, K. Amgain, Corona virus disease (COVID-19) outbreak and the strategy for prevention. Europasian J. Med. Sci. **2**(1), 1–3 (2020)
48. J. Sims Will the world be quieter after the pandemic? (2020). https://www.bbc.com/future/article/20200616-will-the-world-be-quieter-after-the-pandemic
49. N. Singh, Y. Tang, O.A. Ogunseitan, Environmentally sustainable management of used personal protective equipment. Environ. Sci. Technol. (2020)
50. S. Singhal, M. Matto, COVID-19 lockdown: a ventilator for rivers. DownToEarth, ed by M. Somani, vol. 11. p. 100491 (2020). https://www.downtoearth.org.in/blog/covid-19-lockdown-aventilator-for-rivers-70771 (Bioresource Technology Reports)
51. M. Somani, A.N. Srivastava, S.K. Gummadivalli, A. Sharma, Indirect implications of COVID-19 towards sustainable environment: an investigation in Indian context. Biores. Technol. Rep. **11**, 100491 (2020)
52. R. Tripathi, India 21-day lockdown: what is exempted, what is not. 2020. https://economictimes.indiatimes.com/news/politics-and-nation/india-21-day-lockdown-what-is-exempted-what-is-not/articleshow/74798725.cms
53. USEPA . Nitrogen Dioxide (NO_2) Pollution. https://www.epa.gov/no2-pollution/basic-information-about-no2 (2016)
54. N. Van-Doremalen, T. Bushmaker, D.H. Morris, M.G. Holbrook, A. Gamble, B.N. Williamson, Lloyd-Smith J.O. Aerosol and surface stability of SARSCoV-2 as compared with SARS-CoV-1. N. Engl. J. Med. **382**(16), 1564–1567
55. G. Wallace, CNN; 2020. Airlines and TSA Report 96% Drop in Air Travel as Pandemic Continues. https://edition.cnn.com/2020/04/09/politics/airline-passengers-decline/index.html 09 April 2020
56. C. Wang, R. Pan, X. Wan, Y. Tan, L. Xu, C.S. Ho, R.C. Ho, Immediate psychological responses and associated factors during the initial stage of the 2019 coronavirus disease (COVID-19) epidemic among the general population in China. Int. J. Environ. Res. Publ. Health. **17**:1729 (2020)
57. WEF, World Economic Forum; Geneva: 2020. High Noon during Coronavirus Lockdown.https://www.weforum.org/agenda/2020/04/high-noon-lockdown-around-the-world/
58. A.P. Yunus, Y. Masago, Y. Hijioka, COVID-19 and surface water quality: improved lake water quality during the lockdown. Sci. Total Environ. **731**, 139012 (2020)
59. M.A. Zambrano-Monserrate, M.A. Ruanob, L. Sanchez-Alcalde, Indirect effects of COVID-19 on the environment. Sci. Total Environ. **728**, 138813 (2020)

Chapter 7
The Internet of Things and Advanced Applications in Healthcare

Vidhyotma Gandhi and Jaiteg Singh

Abstract Internet of Things (IoT) is a trending technology. The chapter addresses IoT technology, IoT architecture, and IoT advanced applications in healthcare. To prevent patient from bedsores an Electronically Controlled Nursing (ECN) is designed (patent filed). The ECN bed is helpful for bedridden patients to change their positions automatically by hand gesture control. The whole device is self-controlled in specific angles, positions and tilt. The bed has a sanitary trolley under the bed with switch-controlled movement. Hence this bed provides complete patient care and decreases the requirement of the hospital caring staff and all activities of patient can be transmitted to the doctor's room or chamber to update them about the patient's current status/ vital signs in ICU.

7.1 Introduction

IoT is the inter-connection of uniquely distinguishable embedded system. The embedded systems are comprised of considerable number of devices within the prevailing network infrastructure. The idea was proposed by 'Kevin Aushton' in 1999 at Procter & Gamble [1, 2]. IoT technology was developed in 2008/2009. Although the term IoT was used around two decades back but there is lot of research scope still in this field. The "Thing" in this phrase refers to a smart object, a patient with the heart monitoring implant, a bedridden patient with any chronicle disease, a paralyzed patient or a physically disable person, a farm cattle with a RFID chip tag, a farm/house with smart fencing, a vehicle with built-in sensor to alert the driver about the low tire pressure or other ordinary phenomena. Internet of Things is actually a network of networks with unique characteristics [3]. Use of WSN to form mesh of

V. Gandhi (✉) · J. Singh
Chitkara University Institute of Engineering and Technology, Chitkata University, Rajpura, Punjab, India
e-mail: Vidhyotma.gandhi@chitkara.edu.in

J. Singh
e-mail: Jaiteg.singh@chitkara.edu.in

S. L. Fernandes and T. K. Sharma (eds.), *Artificial Intelligence in Industrial Applications*, Learning and Analytics in Intelligent Systems 25,
https://doi.org/10.1007/978-3-030-85383-9_7

network with any type network topology and sending data over network intensifies the efficacy of IoT in all field. In IoT everything is made smart by embedding a unique identification and advance sensors. The sensors' data is made available on internet through mesh networking (wired/wireless) to control the devices or systems [4]. IoT is a pioneering field that integrate multiple traditional technologies to create novel applications. The smart devices are the basic building blocks of IoT those are built with Machine-to-Machine (M2M) communication abilities [5]. IoT provides each object an individual identity by assigning IP address or any other technique and connect it with surroundings. The unique IP address assigned to smart devices makes them able to communicate over network. The gigantic availability of IPV6 addresses plays the vital role in the acceptance of IoT [4].

There is no specific architecture for all types of applications. It is deployed according to application, personal interest and the expertise of person. To make human life more comfortable and safer and secure needs research community attention on IoT standardized architecture, quality of services and communication protocols. This chapter addresses IoT technology, architecture, and advanced applications in healthcare.

7.1.1 Evolution

'Kevin Ashton' during his presentation at 'Procter & Gamble' (P&G) in 1999 first time used the phrase "Internet of Things" by linking the new idea of RFID in the supply chain of the company [2]. At the same time 'Bill Joy' in Davos (Switzerland) in 1999 presented at the 'World Economic Forum' the D2D (Device to Device) communication as part of his "Six Webs" framework which was also a known term within MIT scientists. In the same year the Director of MIT Neil Gershenfeld published his book " When things start to think" [6]. The IoT made possible the interactions between Thing-to-Thing (T2T), Human-to-Human (H2H), Human-to-Thing (H2T), Thing-to-Human (T2H) at a virtual level in day to day life as given in Fig. 7.1. In IoT, smart objects easily communicate with each other, that leads to the vision of "anytime, anywhere, any media and anything" communication [7].

In the IoT, the ***advance healthcare facilities*** is the promising research area. The smart things are able to interact and communicate the environmental data in inter and intra network. The IoT applications, autonomously adapt/react to the real/physical world events and influence them by executing the algorithms/processes that stimulate actions and provide services with/without human involvement [8]. An IoT application is a collection of large number of homogeneous/heterogeneous objects comprised in the process. Wireless communication technologies are used to communicate within objects and end user. Implementation of "Internet of Things" technology can be done in many ways. The technologies used in IoT are discussed in the next section.

Fig. 7.1 Connectivity of IoT

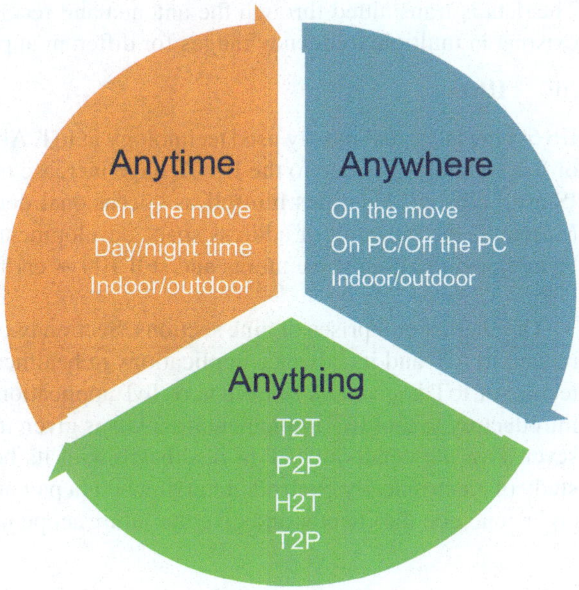

7.2 Technologies Used in IoT

Multiple technologies like RFID, Communication (WSN, WPAN, LAN), IPV4/IPV6 and so on are used to are implemented IoT applications. IoT semantically is the combination of two terms; one is a "network oriented" vision of IoT, and the other emphases on general "objects" to be amalgamated into a general framework. It depends upon developer or user to choose either an "Internet oriented" or a "Things oriented" according to the application or task to be performed, their self-interests, expertise and backgrounds.

Initially, the definition of IoT was derived from a "Things/Objects oriented" perspective; the considered objects were very simple item, "Radio-Frequency-Identification" (RFID) tags/systems [9].

(i) **Wireless Private Area Network (WPAN)**

WPAN can be implemented by using the concept of WSN. Radio Frequency (RF) TX/RX Modules, Zigbee or Bluetooth modules are used to implement WPAN. WSN is a network of any number of nodes. The nodes can be static or mobile and homogeneous or heterogeneous type but should smart and trackable on the network [10].

(ii) **Radio Frequency Identity (RFID):**

In RFID, radio frequency waves are used to recognize/identify a product, individual, substance, animal, car or other object embedded with a tag. RFID technology is the amalgamation of three things: Tag (transponder), Reader (transceiver) and Antenna.

The data is transmitted through the antenna and received by the reader. RFIDs are existing in multiple frequency ranges for different applications [11].

(iii) **IPv6:**

IPv6 is the latest and mostly used technology in IoT. Abrupt development and growth of IoT technology is due to the IPv6's huge increase in address space. According to 'Steve Leibson', identifies himself as "occasional docent at the Computer History Museum," the meaning of address space development is that the IPV6 address can be assigned to all universe atoms and still 100 + earths addresses are left to assign [12].

The chapter comprises of nine sections. Section two depicts the existing literature related to IoT and its advance applications in healthcare. In section four the architecture of IoT is given. A brief idea of IoT applications is given in section four. An introductory text of IoT implementation kits is given in section four. Section six and seven give the details of IoT in healthcare with its benefits & issues and the case study of electronically controlled nursing bed (a patented idea) respectively. Section eight conclude the chapter and give the future scope idea in IoT.

7.3 Literature Review

Lot of research articles related to IoT are published. These articles touched multiple IoT services and applications areas like healthcare, sports, agriculture, logistics, environment, household, industrial and many more [13]. To enhance the capabilities of IoT Common Variability Language (CVL) process was used to create Self-Star MAS agents with basic properties like distributed nature, context-awareness, and self-adaptation. By using domain-independent language, the variability of the heterogeneous and interconnected devices was efficiently resolved [14]. IoT and cloud are trending technologies. IoT devices are small in size so mostly equipped with low storage and computational capacity. On the other hand, cloud is stack of storage and computational capabilities. The complement nature of IoT and cloud boost the capabilities of each other and bring revolution in real time applications development [15]. Many cloud service providers (amazon, Google, Microsoft Azure and few more) are now provisioning the multiple IoT application designing platforms [13]. The blend of Wireless Sensor Network (WSN), Cloud Computing, (CC) and data on demand by web3 technology helped to create a novel integrated Sensor-Architecture-Internet framework. The framework was implemented with the integration of private cloud Aneka and Public cloud Microsoft Azure. Azure allowed Aneka to launch any number of instances. Each instance can have advanced PaaS multiple programming models, runtime execution services, workload management, dynamic provisioning, QoS based scheduling and flexible billing. Aneka was the middleware between Azure and user's applications [16]. A survey of different domains of advancement in IoT-based healthcare like network architecture, industrial trends, applications, data security and privacy features etc. The study focus on IoT and eHealth policies and regulations which are globally adopted by different countries [17]. Few open issues and challenges in the growth of this field are real tie application development [18]. Different IoT applications has different architecture there is no fix architecture that can be followed for all IoT applications. Next section gives an idea of generic IoT architecture.

7.4 Architecture of IOT

IoT architecture's working approach is bottom to top approach. It collects data from sensors on smart object/devices then processes the data in controllers and make decisions. The decisions are communicated to the devices/actuators/persons for further actions. User interface or application layer is in the top most layer. The technologies, assign addresses to smart devices/objects are in middle layer. Data of Smart object's sensors is sent to the processing software's like microcontrollers. Communication protocols used is in the bottommost layer. The layered architecture of IoT is given in Fig. 7.2.

Architecture of "Internet of Things" consists of four layers [19]. The functionalities of each layer are discussed briefly here:

7.4.1 Application Layer

Application layer is the user interface layer present at the highest level of architecture. It is liable for distribution of many applications to diverse users of IoT by establishing sessions. Due to increasing use of RFID in several applications the opportunities of IoT in diverse industry verticals such as healthcare, manufacturing units, environment, logistics, public safety, retail supply chains, food, packaging industry and so on.

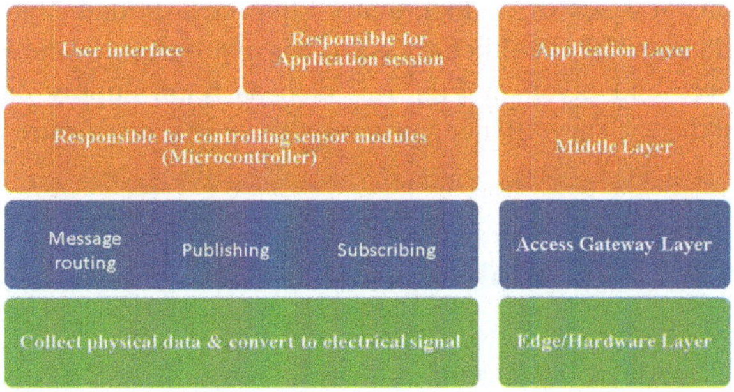

Fig. 7.2 Layered Architecture of IoT

7.4.2 Middleware Layer

Middleware Layer deploys an interface between the top application layer and bottom hardware layer. This layer is liable for analytical functions such as information and device management and semantic analysis. The analytic functions take care of process like data filtering, access control, and information discovery, data aggregation, information service. Middleware layer works in bidirectional mode.

7.4.3 Access Gateway Layer

Access gateway Layer is liable for message routing and schedule publish and subscribe routines. First stage data handling and cross platform communication processes are performed at this layer.

7.4.4 Edge Layer / Hardware Layer

sensor networks, RFID modules, embedded systems and so on forms Edge layer. Edge layer is liable for converting the physical data to electrical data. The primary data is collected by these sensors from the real-time environment. Embedded systems used for information collection, information storage & processing, communication, control & actuation of motors or other devices [19].

7.5 Applications of IOT

The probability of deploying "Internet of Things" in diverse areas like domestic, industry 4.0, retail industry, defense, health sector, education and other numerous business sectors is identically high. Few domains, are currently using it and in few domains like healthcare research is in progress. Some application areas are mentioned here and given in Fig. 7.3 such as retail supply chain, logistics, Manufacturing industry, surveillance, automation, sports, healthcare and many more [20].

7.6 IOT Implementation Kits

Nowadays, in electronics market lots of microcontroller boards are available thar are compatible to design IoT applications. Every controller board is assigned a MAC address for its unique identification in the global network. The boards are capable to

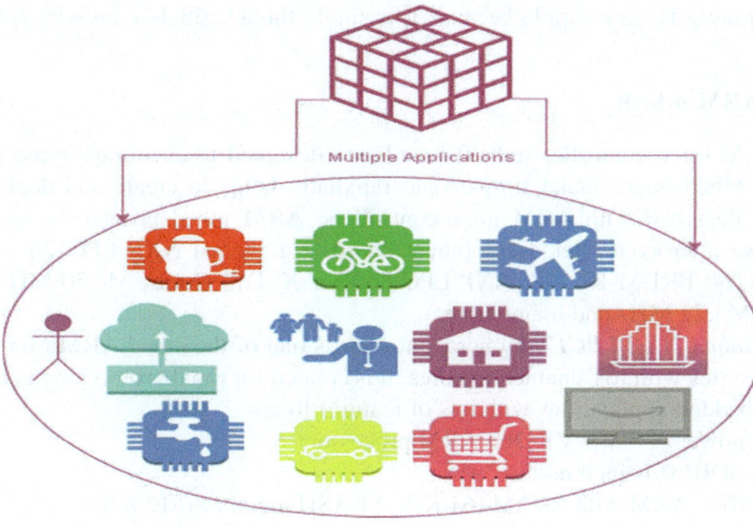

Fig. 7.3 Applications of IoT

handle dynamic IP address that provides the capability to freely communicate in the network. Few IoT capable microcontroller development boards are listed below:

(i) **Arduino**

Arduino provides a series of boards e.g. Arduino Diecimila, Arduino Duemilanove (rev 2009b), Arduino UNO, Arduino Leonardo, Arduino Mega, Arduino Nano, Arduino Due (ARM-based), LilyPad Arduino (rev 2007) with an open source software and hardware. *The features of Arduino UNO* Microcontroller ATmega328 of Atmel used in Arduino UNO. Its operating voltage is 5 V and recommended voltage is 7-12 V. 14digital I/O pins out of which 6 pins feature PWM output. 6 Analog Input Pins. DC Current for I/O Pin is 40 mA. Arduino has 32 KB Flash Memory and Static RAM 2 KB and EEPROM 1 KB. Its Clock Speed is16 MHz.

(ii) **Intel Galileo:**

This board is the first Arduino-certified development boards with Intel processor based on Intel × 86 architecture which can be programmed in any language supported with Eclipse platform. The application processor Intel Quark SoC X1000 and a 32-bit Intel Pentium-class system on a chip is inside the Intel Galileo. Intel Galileo hardware (pin-compatible) and software are compatible with Arduino UNO shields. Digital pins 0 to 13 with adjacent AREF and GND pins, Analog inputs 0 to 5, the power header, ICSP header, and the UART port pins. It has RS-232 Serial Port transceiver, 10/100 Ethernet connector, USB Client connector, USB 2.0 Host connector, 8 MByte Legacy SPI Flash, 7-pin Serial Peripheral Interface (SPI) header, Analog to Digital converter, 512 KBytes SRAM and so on. Intell Gallileo programming is very simple because it is single thread, single core with constant speed.

(iii) **ARM mbed:**

The ARM microcontroller (mbed) boards are designed to encourage rapid prototyping. Mbed board make simpler and rapidiatly helps to create and deploy the devices designed with ARM microcontrollers. ARM mbed provides a series of arm base microcontroller development boards e.g.: mbed NXP LPC1768, mbed LPC11U24, FRDM-KL25Z, NXP LPC 800-MAX, DIP Cortex M0,50 MHz, DIP Cortex-M3,72 MHz and many more.

The ***mbed NXP LPC1768*** microcontroller is one of the mbed ARM Microcontrollers series with IoT enabled features. It is created for rapid prototyping to design IoT/embedded applications with lots of features like:

Low power (4.5 V-9 V) 32bit ARM processor.

40-pinDIP, 0.1- inch pitch.

100 MHz ARM with SRAM (64 KB), FLASH memory (512 KB).

Supports interfaces such as USB, SPI, I^2C, CAN, Ethernet and Serial.

C/C + + cloud based Web compiler.

mbed cloud library access (API driven approach).

4 UARTs and 8-channel DMA controller.

motor control PWM, 12-bit ADC, 10-bit DAC.

Quadrature encoder interface, 4 general purpose timers, 6-output general purpose PWM

(iv) **Raspberry pi Foundation:**

Raspberry pi foundation produces raspberry pi Gen1 and raspberry pi Gen2 with all features like computer on a single handheld board with.

Technical Specifications:

Broadcom BCM2836 Arm7 Quad Core Processor running at 900 MHz.

RAM 1 GB, to run bigger and more powerful applications.

40pin extended GPIO to enhance your "real world" projects.

4 × USB 2 ports.

4 pole Stereo output & Composite video port.

HDMI, CSI camera, DSI display port.

Micro SD port to store operating system and data storage.

Micro USB as power source.

10/100 Ethernet Port Internet connectivity.

Combined 4-pole jack for connecting your stereo audio out and composite video out.

7.7 IOT in Healthcare

IoT is progressively being recognized by investigators and analysts as one of the most promising technologies. IoT have the probable potential to affect the productivity, health and safety of billions of intimates leading to the foremost economic impact. Healthfulness is one of the most prominent application areas of IoT technology. In healthcare, IoT delivers a well-timed cost-effective solution to improve patient care. The recent advancement in sensor fabrication, communication, cloud computing, mobility support and big data technologies have directed towards cost-effective medical devices and associated health programs, immensely intensifying the space of IoT in healthcare and equipment's to take future changes. Conferring to a new report from "marketandmarket", the healthcare IoT market sphere is exalted to hit $118 billion by 2025 [21].

IoT have numerous services scope in the healthcare commerce, with the probability of using the mobile phones embedded with RFID-sensor that have the capability to monitor the medical parameters and drug delivery ratio. Implantable and identifiable wireless devices/sensors can be used to record the health parameters of chronicle patient like heart disease, cancer, diabetes, cognitive impairments, stroke, seizure disorders, Alzheimer's and so on. In future the recorded data can be used to save patient's life in emergency conditions. Paraplegic patients can be delivered muscular stimuli dose through an implanted smart electrical simulation device to reinstate mobility of organs [22]. Following are the main areas of healthcare where IoT can be implemented:

(i) **Pharmaceutical**: The pharmaceutical firms claim to make consumer's life safe. It can be achieved by smart labeling. The technology provides smart labels for drugs which will help in tracking its quality and quantity. Further, aware the forbearing about its dosage's details and storage conditions. For example: stipulate protein dose and analyze accuracy of its composition. It is implemented through device–to–device interaction using big data that is produced by multiple of devices.

(ii) **Medical & Health:** To locate doctor and nurses as well as patient in the hospital at any point any time the wearable sensors are the best solution and vice-versa through IoT applications a message can be communicated to doctors' chamber about the abnormal conditions of patient e.g. BP fluctuation, heart beat rate, pulse rate, glucose level etc. Nurses can be alarmed for medicine timings or any other help needed by the patient. IoT enabled knowledge systems can be used to distinguish adverse effect of drugs in forbearing. The vital bio-parameters of human like sugar, cholesterol, blood pressure, temperature, heart and pulse rate can be recorded and decoded through the combination of sensors, Wi-Fi using IoT technology. Apart from this IoT offer core functionalities needed by clinical, front office and billing staff. Many doctors and customers use the IoT enable core services. For example: Screen hand sterility to reduce the transmission of infections to staff or patients. An important existing application is RFID-labeled blood containers allowing to administrate the time outside the freezer. Otherwise, the time period is unknown and therefore the blood found outside the freezer has to be dumped to eliminate risks.

(iii) **Aid for especially abled or elderly persons:** IoT is identically useful for elderly and especially abled person. The use of wearable wireless sensors with unique identity transmit the health parameters of a person and send alarm for assistance in case of any disorder. IoT enable devices help them to be independent in their daily routine life as "Universal Electronic eye guide for differently abled persons" mentioned in case study. (Project developed and presented in SAP company event "SAPmile DKOM" Feb2016. This project graded with Ist position in this event.)

Few benefits and issues of IoT in healthcare are listed below:

Benefits of IoT in Healthcare:

It lowers the cost of patient care.
It improves the patient's outcomes.
Using IoT real time disease management is possible.
IoT improve quality of life and user experiences.

Issues in IoT healthcare:

Dealing with device interoperability and diversity.
Data integration and privacy.
Scale, data volume and performance.
Flexibility and scaling of applications.
Essential of medical knowledge.
In next section A case study of realtime implementation an electronically controlled nursing bed (a patented project) is discussed.

7.8 Case Study: Electronically Controlled Nursing Bed

The ongoing shortage of medical practitioners and the intensification in the elderly population due to the baby boom after second World War, gears up health care automation industry. Developing countries are continuously struggling to offer better healthcare facilities to its ever-growing elderly population. Precisely, the patient monitoring is very meddlesome and labor-intensive task. Further, it also requires expertise and continuous supervision. According to WHO (World Health Organization) 13% population above the age of 60 years are bed ridden whose scale is increasing day by day and assume to grow about 24% between the years 2020–2023. The average ratio of male and female is 67.5 & 32.5% respectively. Out of them, 89% patients are suffering from unary infection and 11% are facing the problem of bedsores [23]. As per expert opinion, bedridden patients, with bedsores, can recover within 6 to 10 months if they got proper medical attention. But it is not affordable for everyone to get round the clock janitor & medical facilities.

In reference to aforesaid scenario, "smart beds", significantly contribute towards comprehensive smart patient-care environment [24]. The recording and monitoring of essential body parameter is the fundamental requirement of medical care or treatment. The uninterrupted monitoring of essential body vital parameters required the attachment of bio-sensors to the human body in a multiple ways. As per the medical recommendations, the bedridden patients, repositioning schedules of body or the change in sleeping and sitting postures should be on the stretch of two hours and one hour respectively. Research and design of multifunctional smart nursing bed for bed ridden people may substantially contribute towards healthcare solutions [25, 26].

This chapter is aimed to develop an "Electronically Controlled Nursing Bed (ECN bed)" for bed ridden or ICU patient. The ECN bed is equipped with accelerometer, gyroscopic sensors, few push buttons switches and motors. ECN bed enables the patient to smoothly change the position as and whenever he needs or wishes or in the stipulated time. The position changing process is completed with the help of sensors and motors implanted on the ECN bed. This would help to reduce the problem of bedsores as it prevents the occurrence or exacerbation of pressure. It reduces the nursing and caretaker cost and keeps the family tension free. The novelty of the study has been proven by filing of patent in Sep, 2015.

The chapter comprises of seven sections. Introduction was given in Sect. 7.1. The available literature of electronic/smart/nursing bed is stated in section two. After reviewing the literature, research gaps are outlined in section three. Section four presents the problem definition for the current study. The objectives of ECN bed are given in section five. Section six represent the methodology adopted during the design and development of ECN bed. The results and the future scope are mentioned in section seven.

7.8.1 Related Study

The scope of an innovation is portrayed by the means of related literature study. This section offers a review of prominent studies appraising smart nursing solutions for bed ridden patients. Initially, a study was carried out to check the effect of rotating bed on bedsores at bedridden patients in the nursing homes. An inflatable air mattress and tube was implanted on the rolling bed to turn the patient at 15-degree tilted position. An automatic process of patient's position changing from right to left, horizontal and left to right was alternatively set on the interval of 15 min [27]. Researchers also augmented hospital bed through computer assisted system having Electronic Medical Records (EMR) charting capabilities. A "point-of-care" system was designed to display the current position of the patient. The computer system hardware was coupled with the frame hospital nursing bed to fetch the position of patient [26]. Later, researchers also proposed a human-centered design process for hospital beds. It intended to promote patient safety and ease of use. In 2017,

Central bed won the Stanley Caplan User Centered award for providing the patients satisfaction, safety and caretaker efficiency [28]. A threefold hospital bed offering the horizontal fold was also proposed. It did not offer any provision of vertical folding. Lots of mechanics we involved hence it was more complex to operate [29]. Later inventions disclose an automatic side turning nursing bed comprising of a hydraulic elevation rod and an electric control system to control the elevation of the two side plates of bed, so as to ensure that bed ridden patients can turn [30]. Newer inventions also introduced an auxiliary device providing the ambulatory folding and moveable bed, which can also be converted to wheel chair. It is a device that can swap between the bed and wheel chair and can assist in the movement of the patient easily [31]. A few inventions also use a bed mechanism to uplift a lying/sleeping patient at an variable height as well as the position [32]. An automatic care taking system for a bed ridden patient comprising a bed body having three rectangular horizontal bed plates fixed on a bed frame was introduced, it helped the patient to do his/her daily bowl works by themselves. The bed plates can be arranged at different positions [33]. An automatic tri-folding bed with numerous length-oriented supporting plates for mattress was designed. The supporting plates were tilt able and individually can rotate to form a V shaped posture to hold the patient safely [34]. Recent inventions offer an automatic patient turner that automatically and alternately turns an immobile patient from one complete side to the other in a less intrusive manner than manual turning, due to the sequential inflation and deflation of strategically placed inflatable [35]. Later, portable devices were used for rotating bed-ridden patients from their back to their side with little or no assistance from the care provider. The rotation of the patient was with the help of air chambers that function between the patient and bed like support [36, 37].

7.8.2 Problem Definition

The available smart bed techniques offer the standard physiological parameter and the position/posture monitoring. The output of the sensors embedded in the bed permit incessant monitoring of patient movement. According to WHO (World Health Organization) 11% population above the age of 60 years are bed ridden which assume to grow about 24% between the year 2020–2023. Approximately 89% of bed ridden patients are suffering from unary infection [38, 39].

According to experts, these bedridden patients can recover easily within 6 to 10 months if they got proper medical facilities. But it is not affordable for everyone to get the 24 * 7 medical facilities [23].

7.8.3 Objectives

The primary objective of the present invention is to design an electronically controlled nursing bed which can be easily and safely operated by the patient to manage his/her prime needs.

Yet another objective is that the smart nursing bed is able to meet all the care related requirements of the patient such as auto sanitary management, auto sideways turning, room lighting control, room curtain control, calling attendant in case of emergency and also provide the medicine reminder etc.

Other objective of the present invention is to make sure that the nursing bed should be robust and simple enough to provide the compact and lightweight design and all the health care medical facilities. The main objects of bed-making were to avoid impediments by ensuring comfortability and security of patient. To afford physical and psychological comfort and security to the patient.

7.8.4 Proposed Solution

To overcome this type of problem, an Electronically Controlled Nursing (ECN) bed was designed which is controlled by the patient itself. The nursing bed of present invention provides high-quality care to the patient and ensures that the patient should be in a comfortable and in physiologically justified position. Thus, the invention provides an opportunity to increase comfort for the patient and their family and decreases the requirement of the hospital caring staff. The primary aim of the nursing bed is to provide an improved control system where the patient may easily and safely operate the bed to manage his/her prime needs without relying on others.

The project, ECN bed not only provides the horizontal but also provides the vertical positioning of bed to assist the patient for sleeping or resting in left or right position. More comfort to the patient and less requirement of hospital caring staff. Inflation and deflation of balloons were used for turning of patient; thus, the solution was simple and more effective. In addition to this, hand gesture based blanket wrapping and curtain positioning is present providing more comfort to patient. The patient can move up down and sideways also. The whole system acts as an assistive system for a bedridden patient. In other assistance, like adjusting light intensity of room, blanket wrapping, auto sanitary management etc. RTC (Real Time Clock IC) is used for medicine reminder and message is flashed to attendant room on display. The attendant after reading the message comes and gives medicine to the patient. Hence ensuring complete comfort to patient without requiring additional nursing staff. With the help of ballooning mechanism, the patient can move easily to left as well as right.

In this bed, the ballooning technology was used to enhances the possibility of move the patient. Various Inertial sensors provide the automation for curtain control and blanket wrapping system. Auto sanitary is managed and controlled by a switch.

Air pressure controlling system is added in order to inject and remove air in the balloons. RTC (Real Time Clock IC) is used for medicine reminder and message is flashed to attendant room on display. The attendant after reading the message comes to administer the drug to the patient. In addition, one can update relative of patient with every hour or a fixed time of period about the health condition so that, they know all about how the patient is responding.

7.8.5 Methodology

The ECN bed works on the technology of gesture recognition. Most of the features are dependent on the gestures of hands and/or fingers. The basic system consists of.

1. Two inertial sensors- one on each hand for turning patient, curtain positioning and blanket wrapping.
2. Microprocessor and micro controller - Placed near top side of bed or anywhere else according to the feasibility and comfort of patient.
3. Multiple balloons for turning the patient without jerk.
4. One switch board (remote with light, fan, trolley call and attendant call)
5. RTC for medicine reminder

During the implementation of ECN bed, the whole process was divided in five parts: (i) Design of bed, (ii) positioning of balloons, (iii) mechanism for turning the patient, (iv) mechanism for blanket wrapping, (v) mechanism for auto sanitary support, (vi) mechanism for medicine reminder and room environment control.

(i) **Design of bed**

The electronically controlled nursing bed comprises of 3 basic parts as shown in Fig. 7.4:

Base body of the bed
Central portion for lifting or turning the patient
Slot for sanitary equipment placement.

(ii) **Positioning of balloons**

Referring to Fig. 7.5, B1, B2, B3, B4, B5, B6, B7, B8 and B9 were the balloons at different positions that aid in the movement of patient. For patient turning fusion of bed lifting and ballooning technology was used that tilts the patient without any jerk in a smooth way. B1 and B2 were the balloons used for turning and supporting the patients. B4 and B5 were cylindrical shaped, present beneath B1 and B2 and were used for supporting the balloons B1 and B2 respectively. B3 was present at the central part of the bed operated by hydraulics (H5) for the vertical movement of the central part. B6, B7 and B8 were the balloons present at the toe end for lifting up the legs of patient when needed. B9 was the balloon at the head that facilitates in lifting the head up during turning.

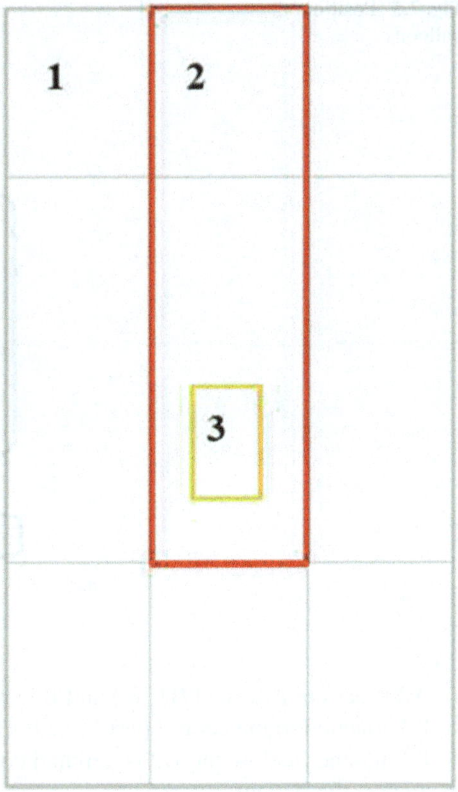

Fig. 7.4 ECN design

(iii) **Mechanism for turning the patient**

The accelerometer sensors were used for turning patient, curtain positioning and blanket wrapping. Two sensors, one on each hand, are used for the above said three functions. When the patient tilts his left hand, he gets turned on his right side and by tilting right hand, he can turn himself on right side. The hands need to be turned in opposite directions i.e. palm side up i.e. Left hand should be turned in right direction with palm and fingers turned on right.

Balloons on various locations help him in taking these left or right sides, without a jerk by the following method.

During the initial condition when the patient is ready to be turned, the balloons B1, B2, B5 and B4 are inflated completely to support the patient while lifting up the central part of the bed.

B3 is deflated at this time.

The hydraulic arm (H5) lifts up the central part with the aid of rod (R).

The balloons B1, B3 and B5 are completely inflated when the patient is in midway during turning.

B2 and B4 are deflated slowly so as to turn the patient.

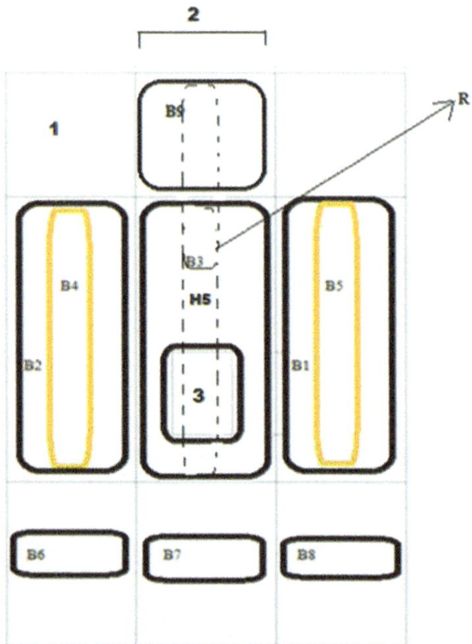

Fig. 7.5 Positioning of balloons

After patient is turned B1, B3 and B5 are fully inflated.

B3 attached to the central part of bed is turned thus tilting the patient.

B4 attached below the B2 is deflated completely and B2 is inflated a little so as to support the patient.

(iv) **Mechanism for blanket wrapping**

The electronically controlled nursing bed of present invention provides complete care to the patient viz blanket wrapping, curtain positioning, light adjustment, auto sanitary functions etc. For blanket wrapping the inertial sensors are used. The whole mechanism is shown in Fig. 7.6. For automatic blanket wrapping hydraulics, pulleys and fiber wires are used. H1, H2, H3 and H4 are the four hydraulics near the four corners of the bed to lift the blanket up to a certain height so that it does not cause any discomfort to the patient. The pulleys P1, P2, P3 and P4 mounted on hydraulics to pull blanket to cover or uncover the patient. The Fiber wires F1, F2, F3, and F4 are attached to the pulleys and the blanket to pull the blanket accordingly. F1 is attached to P1 and F2 is attached to P2 are used for blanket wrapping. F3 with P3 and F4 with P4 inside the blanket beading are used for blanket unwrapping.

(v) **Mechanism for auto sanitary support**

The Auto- Sanitary management was based on trolley system as shown in Fig. 7.7, that was controlled by switch. The patient can press the switch that send intimation to the caregiver to bring the trolley near the bed. The excretion system comprises of a urinary pan (5) for toileting of a bed ridden patient. It is attached to the stand

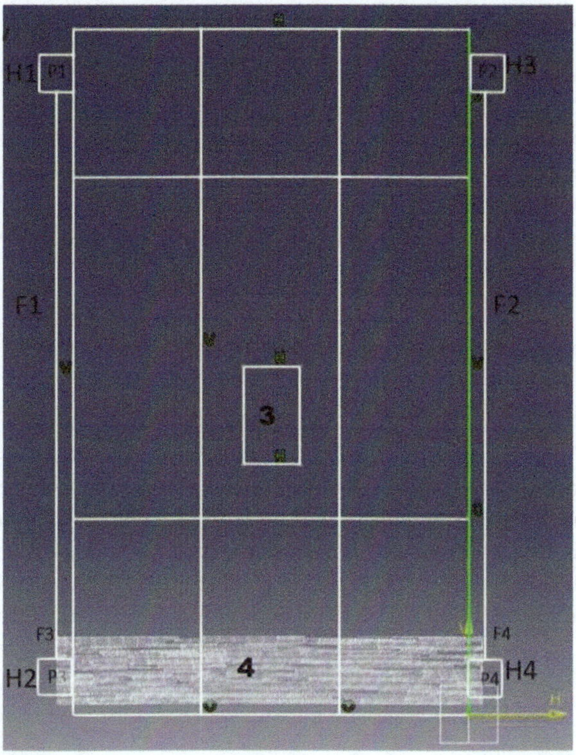

Fig. 7.6 Blanket wrapping mechanism

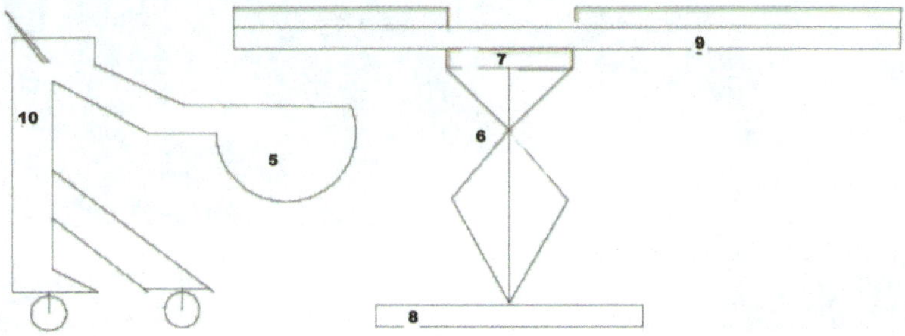

Fig. 7.7 Auto sanitary mechanism

(10) for easy maintenance and maneuverings. The pantograph scissors mechanism (6) is used for lifting and lowering the bed part (7) for excretion. The pantograph mechanism used would depend on hydraulics. Caretaker/attendant can help patient in his/her bowel movement without lifting the patient form bed.

(vi) **Mechanism for medicine reminder and room environment control**

RTC (Real Time Clock IC) was used for Medicine reminder and message was flashed to attendant room on 16X2 LCD (Liquid Crystal Diode) display. The attendant after reading the message came to administer the drug to the patient. In addition to this a switch-controlled call to attendant facility was provided to patient, so that he may call the attendant anytime.

The patient can ON or OFF the fan and light by its own with the help of switches or remote provided on the bed.

7.8.6 Results

The successful implementation of ECN bed was done. The glimpses of implementation process are shown in Fig. 7.8. The fusion of bed base lifting and ballooning technologies made the patient able to turn on the either side, right or left without any jerk in a smooth way. Various sensors and switches are provided on the electronically nursing bed that assists the patient to manage well his daily chores without relying on the caretaker or any care taker. The ECN bed finds a varied industrial application in

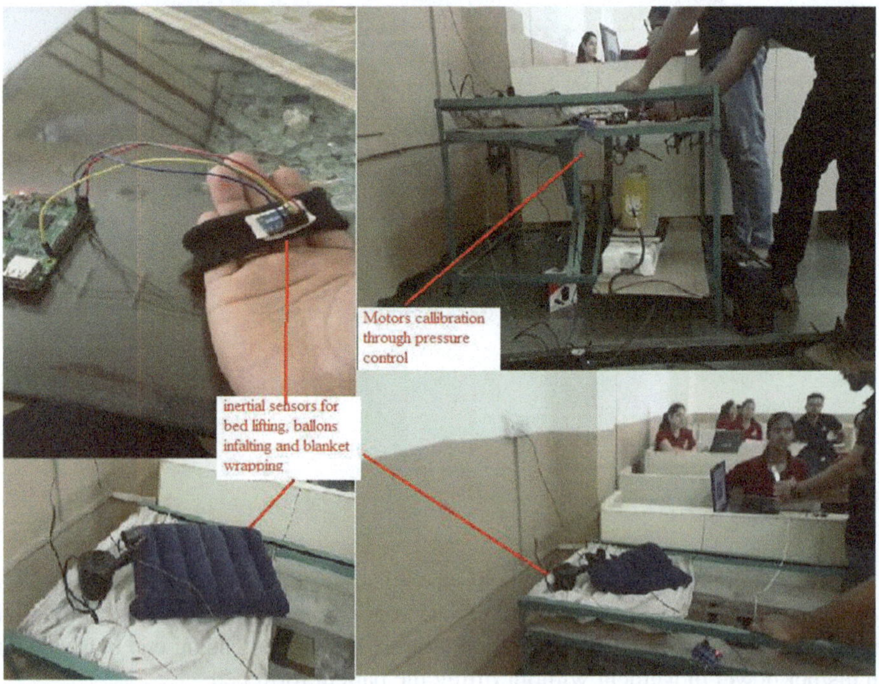

Fig. 7.8 Implementation process

patient care systems and hospitals (Intensive Care Unit). The nursing bed of present invention acts as a standalone system in patient care as the patient was himself able to adjust the settings of bed without the help of caregiver. The nursing bed was simple, compact, lightweight and was robust and stable.

7.9 Conclusion Future Scope

The IoT is a promising technology and have wide scope in vertical applications especially in advance healthcare solutions. The case study ECN bed finds a varied industrial application in patient home and hospital care systems. The nursing bed acts as a standalone system in patient care as the patient is himself able to adjust the settings of bed without the help of caregiver. The nursing bed is simple, compact and lightweight making it robust and stable. Similar type of more patient caring systems such as paralysis patient caring chair can be designed by using the same technology.

References

1. A.L.P.A. Dinesh Bhatia, D. Bhatia, A.L.P. Aroul, Antenna designs for wearable body sensor communication. ACM Trans. Sens. Networks, **10**(4), 1–27 (Mar. 2013)
2. Kevin Ashton, That 'Internet of Things' Thing - 2009–06–22 - Page 1 - RFID Journal. RFID J. (2009). [Online]. Available: http://www.rfidjournal.com/articles/view?4986. Accessed 14-Dec-2017
3. P.P. Ray, A survey on internet of things architectures. J. King Saud Univ. - Comput. Inf. Sci. **30**(3), 291–319 (Jul. 2018)
4. S. Madakam, R. Ramaswamy, S. Tripathi, Internet of Things (IoT): a literature review. J. Comput. Commun. **3**(3), 164–173 (2015)
5. Geng Wu, S. Talwar, K. Johnsson, N. Himayat, K.D. Johnson, M2M: from mobile to embedded internet. IEEE Commun. Mag. **49**(4), 36–43 (Apr. 2011)
6. J.P. Page, ETC: Bill Joy's Six Webs | MIT Technology Review. Page, J. P. (no date) ETC: Bill Joy's Six Webs | MIT Technology Review. Available at: https://www.technologyreview.com/2005/09/29/230292/etc-bill-joys-six-webs/. Accessed 3 June 2021. [Online]. Available: https://www.technologyreview.com/2005/09/29/230292/etc-bill-joys-six-webs/. Accessed 03-Jun-2021
7. G.M. Luigi Atzori, Antonio Iera, L. Atzori, A. Iera, G. Morabito, The internet of things: a survey. Comput. Networks **54**(15), 2787–2805 (Oct. 2010)
8. H. Sundmaeker, P. Guillemin, P. Friess, S. Woelfflé, Vision and Challenges for Realising the Internet of Things (2010)
9. G. Kortuem, F. Kawsar, V. Sundramoorthy, D. Fitton, Smart objects as building blocks for the Internet of things. IEEE Internet Comput. **14**(1), 44–51 (Jan. 2010)
10. A. Milenković, C. Otto, E. Jovanov, Wireless sensor networks for personal health monitoring: issues and an implementation. Comput. Commun. **29**(13–14), 2521–2533 (2006)
11. E.W.T. Ngaia Karen, K.L. Moonb Frederick, J. Rigginsc Candace, Y. Yi, E.W.T. Ngai, K.K.L. Moon, F.J. Riggins, C.Y. Yi, RFID research: an academic literature review (1995–2005) and future research directions. Int. J. Prod. Econ. **112**(2), 510–520 (Apr. 2008)
12. J.H.T. Winter, Ed. P. Thubert, Ed, A. Brandt, RPL: IPv6 routing protocol for low-power and lossy networks [IETF RFC 6550]. Kenchiku Setsubi Iji Hozen Suishin Kyōkai (2014)

13. P.P. Ray, A survey of IoT cloud platforms. Futur. Comput. Informatics J. **1**(1–2), 35–46 (Dec. 2016)
14. I. Ayala, M. Amor, L. Fuentes, *Towards a CVL Process to Develop Agents for the IoT* (Springer, Cham, 2014), pp. 304–311
15. A.P. Muhammad Younas, Irfan Awan, Internet of things and cloud services. Futur. Gener. Comput. Syst. **56**, 605–606 (2016)
16. J. Gubbi, R. Buyya, S. Marusic, M. Palaniswami, Internet of things (IoT): a vision, architectural elements, and future directions. Futur. Gener. Comput. Syst. **29**(7), 1645–1660 (Sep. 2013)
17. S.M.R. Islam, D. Kwak, H. Kabir, M. Hossain, K.-S. Kwak, The internet of things for health care : a comprehensive survey. IEEE Access 2015 **3**, 678–708 (2015)
18. D. Glaroudis, A. Iossifides, P. Chatzimisios, Survey, comparison and research challenges of IoT application protocols for smart farming. Comput. Networks **168**, 107037 (Feb. 2020)
19. D. Chen, G. Chang, L. Jin, X. Ren, J. Li, F. Li, A novel secure architecture for the internet of things. In 2011 Fifth International Conference on Genetic and Evolutionary Computing, pp. 311–314 (2011)
20. D. Zeng, S. Guo, Z. Cheng, The web of things: a survey. J. Commun. **6**(6), 424–438 (2011)
21. IoT Healthcare Market Size, Share and Global Market Forecast to 2025 | COVID-19 Impact Analysis | MarketsandMarkets, *Report Code: TC 3847*. [Online]. Available: https://www.marketsandmarkets.com/Market-Reports/iot-healthcare-market-160082804.html. Accessed 14-Jun-2021
22. J. Gómez, B. Oviedo, E. Zhuma, Patient Monitoring System Based on Internet of Things. Procedia Comput. Sci. **83**(Ant), 90–97 (2016)
23. P. Bains, A.S. Minhas, Profile of home-based caregivers of bedridden patients in North India. Indian J. Community Med. **36**(2), 114–119 (Apr. 2011)
24. I. Ghersi, M. Mariño, M.T. Miralles, Smart medical beds in patient-care environments of the twenty-first century: a state-of-art survey. BMC Med. Inform. Decis. Mak. **18**(1), 1–12 (Jul. 2018)
25. G. Zhan, J. Lyu, M. Chen, Research and Design of multifunctional nursing bed for Elderly People. In IOP Conference Series: Mater. Sci. Eng. **394**(3), 032131 (2018)
26. R.M.Z.A.R.D.V.E.C.W.R.D. Wildman, US10176297B2 - Hospital bed computer system having EMR charting capability - Google Patents (2019)
27. T. Izutsu, T. Matsui, T. Satoh, T. Tsuji, H. Sasaki, Effect of rolling bed on decubitus in bedridden nursing home patients. Tohoku J. Exp. Med. **184**(2), 153–157 (1998)
28. N. Wiggermann, K. Rempel, R. M. Zerhusen, T. Pelo, N. Mann, Human-centered design process for a hospital bed: promoting patient safety and ease of use. Ergon. Des. Q. Hum. Factors Appl. **27**(2), 4–12 (Apr. 2019)
29. Improved three-fold hospital bed (22-Dec-2000)
30. T. Yamamoto, Automatic side-turning nursing invalid bed (30-Sep-2001)
31. Sickbed patients nursing assistant device (15-Sep-2014)
32. Europaisches Patentamt European Patent Office (Dec. 1988)
33. Romana FrancesconiStefano Perfetti, Bed assembly to lift up a lying patient at an adjustable height and position (06-Sep-1998)
34. T. Yamamoto, United States Patent (19) AUTOMATIC CARE-TAKING SYSTEM IN USE FOR BED-REDDEN PATENT (Nov. 1990)
35. J.P.C.L.M.J. Meyer, Automated tri-fold bed (Sep. 1996)
36. M. Ben-Levi, U. S. Patent US007080422B2 Automatic patient turner (Feb. 2006)
37. B.K. Townsend, Inflatable device for turning people on their side and back again (Jul. 2006)
38. *WORLD HEALTH STATISTICS 2 0 1 0*. 2010
39. WHO | World Health Organization. [Online]. Available. https://www.who.int/gho/mortality_burden_disease/en/. Accessed 10-Jun-2020

Chapter 8
A Comprehensive Review of Recent Automatic Speech Summarization and Keyword Identification Techniques

Tapesh Kumar, Mehul Mahrishi, and Gaurav Meena

Abstract Speech has been the most popular form of human communication. A keyboard or a mouse, on the other hand, is the most common way of entering data into a computer. It would be wonderful if computers could understand and carry out human commands. The method of obtaining the transcription (word sequence) of an utterance from the speech waveform is known as automatic speech recognition (ASR). Over the last few decades, speech technology and systems in human-computer interaction have progressed progressively and significantly. This chapter suggests a comprehensive review of automatic speech recognition systems (ASR) and their most recent developments. This research aims to outline and explain some of the popular approaches in speech recognition systems at various stages and highlight selected systems' unique and innovative characteristics.

Keywords Automatic speech recognition · Keyword search · Machine learning · Text summarization

8.1 Introduction

As Internet applications proliferate, extraordinary volumes of text documents and multimedia content have grown significantly. But unfortunately, this vast information leads to information oversupply, making essential knowledge inaccessible. Several applications for an image to text, speech to text, text summarization, and keyword identification prevail in the research community.

T. Kumar · M. Mahrishi (✉)
Swami Keshvanand Institute of Technology, Jaipur, India
e-mail: mehul@skit.ac.in

G. Meena
Central University of Rajasthan, Ajmer, India
e-mail: gaurav.meena@curaj.ac.in

S. L. Fernandes and T. K. Sharma (eds.), *Artificial Intelligence in Industrial Applications*, Learning and Analytics in Intelligent Systems 25,
https://doi.org/10.1007/978-3-030-85383-9_8

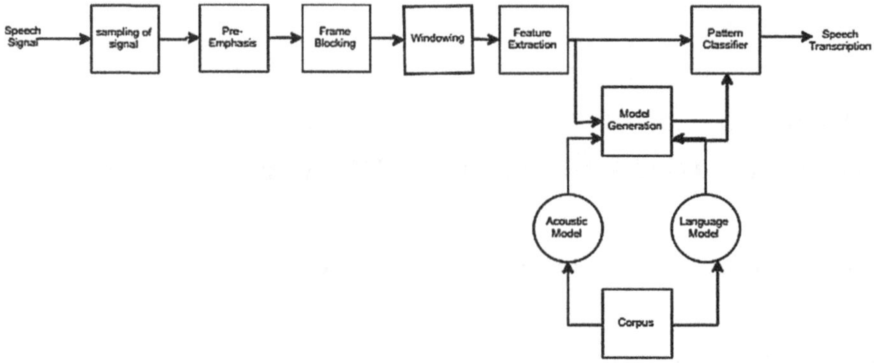

Fig. 8.1 Block Diagram of ASR

The use of words is the most potent form of communication through which people can convey their ideas and emotions. When it comes to the characteristics of speech, each language has its uniqueness. The rate and dialect, however, differ while speaking in the same language. As a result, people who struggle to grasp the message's meaning find it challenging to do so. In addition, speeches with long pronunciations may sometimes be hard to follow since various speakers might pronounce words differently. Through this chapter, we investigate the state-of-the-art methods for automatic keyword detection using ASR techniques.

Figure 8.1 showcases the block diagram for a basic ASR. Digital speech is the input to the system in the form of signals. These signals undergo a series of steps before a pattern is recognized from the speech signals.

- **Speech Sampling**: is a strategy for increasing recognition accuracy. To ensure accurate sampling, it is essential to have a higher sample frequency. Commercial voice recognition systems often utilize analog values of similar parameters, and the results are generally excellent.
- **Pre-emphasis**: The dynamic range of the digitized speech waveform is large, and it is vulnerable to additive noise. As a result, pre-emphasis is employed to flatten the signal's spectral profile, making it less sensitive to finite precision effects during speech processing. The fixed first-order technique is the most widely used pre-emphasis method.
- **Frame blocking**: The voice signal is time-variant or dynamic; hence frame blocking is necessary. When a voice signal is evaluated for a short time, it is believed to be stationary.
- **Windowing**: Each frame is windowed to decrease signal discontinuities at the frame's beginning and ending. Spectral distortion is minimized by using the window to taper the signal to zero at the beginning and end of each frame.
- **Corpus**: is a dataset of machine-readable texts generated in a natural communicative context. It represents that language variety [24].

- **Language Modelling**: is the single most complex component, having been trained on millions of words and containing millions of parameters. It was created to recognize the links between words in a phrase using a pronunciation dictionary. ASR systems employ n-gram language models to help them find the correct word sequence by evaluating the likelihood of the nth word based on the n-1 previous words.
- **Acoustics Modelling**: is an essential component of an ASR since it accounts for much of the computational load and performance of the system. It's utilized to connect the speech signals' observance aspects with the hypothesis sentence's expected phonetics. The Acoustic model is developed for detecting the spoken phoneme [25].
- **Pattern Classification**: It means developing a function that translates the input feature space to an output space with two or more classes. A neural network (NN) is an effective pattern categorization tool that builds a model using training and testing data.

8.2 Brief History of Automatic Text Summarization and Keyword Extraction

Many researchers have developed frameworks and systems for video segmentation either for information retrieval or for content-wise indexing. [1] is one of the oldest pieces of literature in this field that use a combination of content analysis and physical property data to pick the shots shown to users. [27] proposes a novel algorithm *'TopicTiling'* which is based on the standard *'TextTiling'* algorithm for text segmentation where blocks of text are compared via bag-of-words vectors. Their work proves to be significant in terms of complexity and computationally less expensive than other LDA-based segmentation methods. [28] research hyper videos that introduce hyperlinks within a video. The work focuses on the educational domain and provides clickable access to video components like pictures, webpages, and text [6]. MMToC created a keyword ranking system. The words were extracted from visual displays as well as spoken using speech transcripts. Based on the ranking words, a cost function is developed. Finally, the topic-wise indexing of the movie is generated using the dynamic Programming optimization approach. Segmentation of a lecture video based on linguistic characteristics of the text. [20] collects and compares text samples from lecture videos when the text-boundary similarity is limited. The major challenge that videos indexing researchers have is distinguishing between the movie's context, i.e. the effect of the film, and the description, which pertains to the actual content of the video. [33] aims to provide a digital index to education videos by converting videos into images and extracting the text directly through the Google Tesseract OCR Engine. [22] is one step ahead of [33]. The concept of both types of research is the same, unlike the latter extracts and recognizes the text directly from the video rather than converting the whole video into images. [2] worked on lecture webcasts and tried to provide a searchable text index so that users can access local material within a video. The technique focuses on videos with presentation slides only and performed key-frame extraction using associativity of the textual content. [14] . Proposed a new way to improve the interpretability and manipulability of a video by combined key-frame extraction and object-based video segmentation method. The similarity and redundancy between the frames are removed through the Kullback Leibler distance (AIKLD) criterion.

Through this extensive literature survey, we can conclude that vigorous research is ongoing in automatic indexing, annotation of videos, feature extraction, video classification, synchronization of video scenes, change in the scene in a dynamic video and concept detection. However, there is a severe paucity of research on automatic video annotation and indexing of educational movies using Deep Learning techniques [30].

8.3 Automatic Video Indexing using Deep Learning

Deep Learning for automatic video indexing opens a whole new dimension. There are numerous research articles and proposed ideas of deep learning techniques, which are already used for video indexing.

A method of automated indexing of videos using a text-based query is suggested by Vijaya Kumar Kamabathula and Sridhar Iyer [17]. The users are asked to provide the keywords as a query; the framework returns a list of videos. S Jothilakshmi [15] worked on the NEWS Dataset and presented a strategy for automated scene content recognition. The drawback of this method is that transcriptions are acquired using commercial technologies, and news scenarios are classified into one of six predetermined categories (National Politics, National News, World, Finance, Society & Culture, and Sports) [21]. [37] offers a technique for collecting video text panoramas that use natural scene text placement, text tracking, and picture stitching to analyze video text information automatically. It was discovered that [35] got a very significant 95.3 percent Recall and 98.6 percent Precision while utilising CNN with Cosine Similarity. Deep convolutional neural networks coupled with OCR tools are currently utilised to identify and recognise video text.

8.4 Taxonomy of Automatic Speech Recognition Techniques

Figure 8.2 shows that there are 3 main approaches in ASR. They are as follow:

1. Pattern Recognition Approach
2. Acoustic Phonetic Approach
3. Artificial Intelligence (AI) Approach

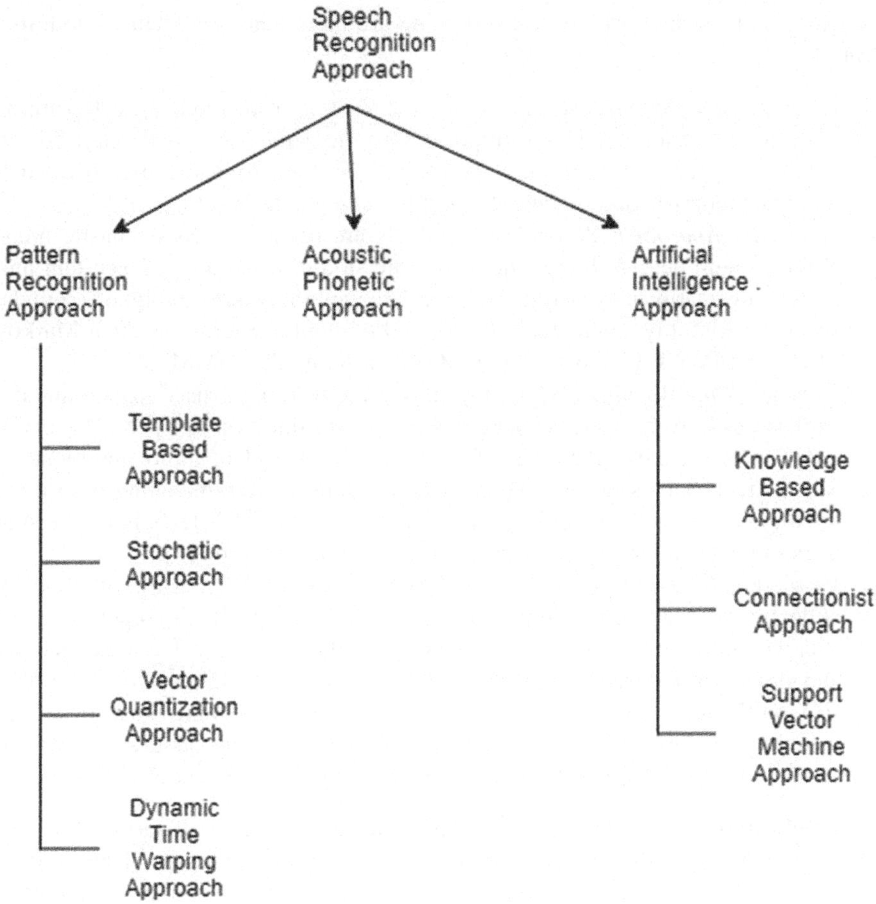

Fig. 8.2 Taxonomy of automatic speech recognition techniques

Acoustic Phonetic Approach: The acoustic-phonetic method [5] proposes that in spoken language, there are distinct phonetic units that are defined by a set of acoustics features. The acoustic properties of phonetic units change depending on the speakers and the surrounding noises (co-articulation effect). This method assumes that the principles regulating variability are simple and can be learned by a computer. It is divided into four stages: speech spectrum analysis, feature extraction, segmentation, and labelling. The acoustic-phonetic method has not been extensively adopted.

Pattern Recognition Approach: Pattern-matching [16] is a mechanism for extracting patterns based on certain criteria and distinguishing one class from another. Feature measurement, pattern training, pattern classification, and decision logic are the four steps. Pattern recognition has been the most prevalent approach for voice recognition in recent decades. The following are the methods associated with this approach:

i. *Template Based Approach*: In this method, a set of prototypical speech patterns is saved as a reference pattern representing the candidate's dictionary. For an unknown speech, recognition is performed by comparing the speech to each reference template and picking the best matching pattern category [29].
ii. *Stochastic Approach*: Probabilistic models are often used in stochastic modelling to help deal with inadequate and uncertain information. Uncertain and partial information may be found in speech recognition due to ambiguous sounds, speaker variability, contextual effects, and homophone terms. Hidden Markov Model (HMM) is the most often used stochastic model (HMM) [5].
iii. *Dynamic Time Warping (DTW)*: Dynamic time warping is used to examine the similarities between two sequences that vary in duration or pace. The DTW allows you to explore many data formats, including video, audio, and pictures. This system enables the best sequence match. Some pattern-matching algorithms are more concerned with consistency than DWT. The DTW technique is best at filling in gaps in lengthy sequences with partial information [4].
iv. *Vector Quantization (VQ)*: A codebook is constructed using this method by collecting each speaker's training characteristics individually. An input Voice is a vector quantified at the reconnaissance stage by the speakers' reference codebook and is used for decision-making to deform the whole sample from the VQ. VQ, on the other hand, lacks temporal information. VQ is the most successful technique for text-independent recognition. Because it makes effective use of small codebooks, VQ is helpful for reference models and codebook searchers [4].

Artificial Intelligence (AI) Approach: The AI technique is a combination of pattern recognition and acoustic-phonetic approaches. It integrates Acoustic Phonetics concepts as well as pattern recognition approaches. Following is some of the methods associated with this approach:

i. *Knowledge Based Approach*: Knowledge-based methods make use of linguistic, phonetic, and spectrogram data. Research has shown that acoustic-phonetic identification systems have been built to identify speech sounds. Voice analysis is integrated with knowledge engineering to provide expert knowledge. But the mixed bag of research findings found when using this method makes it challenging to bring together information from both an expert and a human perspective. While knowledge-based systems may use one of many speech unit types and recognize spoken speech, they are most useful in the selection of a suitable speech unit and the development of a recognition algorithm [4].
ii. *Connectionist Approach (ANN)*: The ANN or Neural Networks is sometimes referred to as a connectionist model (NN). Having good training and learning techniques readily available is very important for these models. Non-linear classifiers that are very sophisticated may be created using Multi-layer Neural Networks that have many layers. This artificial intelligence approach utilizes lexical and syntactic information and technology like neural networks to learn and categorize pho-

netic units like syllables and segments for labelling and segmentation. Due to the simplicity and uniformity of the processing components, hardware implementation of the connectionist paradigm is advantageous [4].

iii. *Support Vector Machine (SVM)*: SVM is a discriminative pattern recognition tool that might be regarded as one of the most powerful technologies available. For data classification, SVM employs both linear and non-linear hyperplanes. The disadvantage of the SVM approach is that it can only classify fixed-length data. It cannot classify variable-length data. However, Before being given to the SVM for classification, variable-length data can be transformed to fixed-length data. It's a generalized linear classifier with a maximum margin fitting function that enables regularisation, allowing it to generalize better. By adjusting the VC dimensions of its models, the SVM controls the model complexity [4].

8.5 Speech summarization and Keyword Extraction based on Neural Network

Kamabathula et al., in their study, provided aims at building a system that, when given a collection of search terms. Following features are packed in the research:

- A list of video readings containing keywords
- Highlight the portion of each video in which the keyword appeared
- Direct links to presentation starting location

To find lectures for specific phrases, users enter the search terms into the search box. This results in a list of lecture videos, which is based on the keywords entered. When keywords occur in any portion of a video, parts of that video are marked. The video in the user interface's media player starts to play when a user clicks on a particular section. The programme includes connections to transition slips, slides now shown, and the speech recording, among other media features. Any word in the voice transcript may be used to go to the most recently played video. Now, only English-language lecture videos use this approach. They had an average accuracy of 0.84 and a recall of 0.72. Despite achieving an average word recognition accuracy of 45 per cent in the voice transcription technique, the decline in recall and precision outcomes was influenced by this low value of recognition accuracy [17].

Research conducted by Hossein Hadian et al. suggests an end-to-end training in the framework of Hidden Markov Models based on the goal function of the maximal reciprocal lattice-free maximum information (LF-MMI). End-of-training refers to training a single deep neural network (DNN) in a single step without using previously learned models, forced alignments, or the construction of state-tying decision trees. They employed complete phones to enable context-aware modelling without the need for trees. The DNN outputs in standard LF-MMI correspond to tied bi-phone or triphone HMM states, where the tying is done via a context-dependency tree. This tree is then built utilizing alignments from an HMM-GMM system. They remove this prerequisite by using monophonic or full biphones. Furthermore, instead of the specific acyclic graph used in standard LF-MMI, they employed the composite HMM (with self-loops) as the numerator graph. As a result, unlike conventional LF-MMI, there is no preliminary alignment information in the numerator graph, and there are no restrictions on the self-loops, giving the neural network considerably more freedom to learn the alignments. Because the alignments for the training data did not exist, the denominator graph telephone language model was generated by utilizing the workout transcripts after quiet 0.2 between words and 0.8 at start and end of sentences was added. The HMM transition probabilities are fixed in this setup. There is no state tying, so the neural network output can fully replicate their effect; training these should not make a difference. This acoustic model is all-neural, tree-free, GMM-free, and is training at one stage flat-start without any previous estimations, initial alignments, pre-workout, or transition training, with a lattice-free MMI. This model is a flat-starter system. These results are like standard LF-MMI outcomes for well-known, large-scale vocabulary assignments. For example, there is a 5 to 25 per cent relative WER reduction on various large and small vocabulary tasks than other end-to-end methods such as Connectionist Temporal Classification (CTC), RNN-Transducers and attention-based methods. On the 2000hr Fisher+Switchboard database, this end-to-end model achieves a WER of 11.0 on all eval2000 (7.5 on the Switchboard subset) in the phoneme-based scenario, and a WER of 12.8 on all eval2000 (8.0 on the Switchboard subset) in the character-based case [12].

Afouras et al. performed research that focused on recognizing words and phrases said by a talking visage, regardless of whether they were accompanied by audio. Rather than using previous attempts to detect a limited collection of words or phrases, they treat lip reading as an open-world issue, encompassing everything from unrestricted natural language sentences to videos in the wild. They investigate how much lip-reading complements speech recognition on noisy audio data.

For the CTC-based or seq2seq method, recurrent neural networks have historically played a large role, but this trend has recently changed. Instead, convolutional neural networks, which focus only on convolutional operations, such as completely convolutional networks for ASR with CTC, are often used. This explains why the TM-seq2seq model works better at lip-reading concerning WER. While the two methods work similarly for audio-only or audio-visual activities, they differ for activities with a video component. TM-CTC models do a much better job in noisy environments; nevertheless, compared to the audio-only and audio-visual TM-seq2seq models, the TM-CTC models perform significantly worse [3].

A neural network is used for a LAS system to collect acoustic, pronunciation, and language modelling data and connect it to an ASR system that relies on speech recognition components. Until now, designs proven to be comparable to state-of-the-art ASR systems on dictation tasks were assumed to be feasible for more complex tasks like voice search, but it was unclear whether this design would be practical. To find structural and optimization improvements that would result in a significant performance improvement. This suggests that words made up of many parts, not individual letters, may be used on the structural side. They also provide a multi-head attention architecture, which performs better than the single-head attentional architectures often used. These techniques have all been shown to lead to better accuracy in the long run: Synchronous training, planned sampling, label smoothing, and minimum word error rate optimization. Finally, they used unidirectional LSTM encoders to show streaming recognition results. While looking into potential improvements, the researchers found that the suggested changes increased the WER from 9.2 to 5.6% on a 12,500-hour voice search job, while the best conventional system obtained 6.7%, concerning dictation tasks [8].

Recurrent neural networks (RNNs) are increasingly being used in the latest voice recognition software to increase resilience. For this reason, the long-term speech and temporal contexts of these voices are entirely genuine. One of the most famous RNN models, gated recurrent units (GRUs), was recently modified by Ravanelli et al., and a simpler, more basic model was then suggested for ASR. The research consists of two phases: The experts analyze the reset gate and show significant redundancy in their function with the update gate. The recommendation was that the old GRU component is eliminated from the overall GRU design, leading to a more efficient and compact single-gate component. To assist with training, they suggested replacing hyperbolic tangent activations with rectified linear unit activations. Results indicate that the suggested design, called GRU, lowers training time by more than 30% each epoch while reducing errors across a variety of tasks, noisy circumstances, and ASR paradigms [26].

Automatic speech recognition systems use SAD as a crucial component for performance. The research by Gelly et al. aims to see whether RNN-based speech activity detection can be improved. Using this strategy, system settings, including feature extraction parameters, NN weights, and backend parameters, are optimized. Three different cost functions are considered when applying SAD optimization: the frame error rate, the NIST detection cost function, and the downstream speech recognizer's word error rate. Models and optimization methods for RNNs are explored in many ways. A trio of different RNN models are explored: a basic RNN, an LSTM with peepholes, and the Gelly and Gauvain LSTM (CG-LSTM), which includes interleaved gates.

When nondifferentiable optimization issues need to be handled, the feature extraction, posterior smoothing, and the initial training of neural networks are improved using a quantum-behaved particle swarm optimization approach. To show experimental SAD results, the NIST 2015 SAD assessment data and the REPERE and AMI meeting corpora are utilized. Results provided utilizing OpenKWS'13 test data also include speech recognition. SAD performance for all activities and circumstances increases because of this modification, and the CG-LSTM model performs better than all other SAD methods tested [11].

The Sequence-to-Sequence ConvErsion Neural Network of Zhang et al. was the first neural network for acoustic modelling in speech conversion (SCENT). SCENT model training is carried out utilizing an attention mechanism that matches the feature sequences of the source and target speakers. The unified acoustic model is used to transpose source utterances' acoustic properties and durations. Mel scale spectrograms, which provide both excitation and vocal tract descriptions of speech signals, are used to analyze acoustic properties.

The source speech bottleneck characteristics are connected as an auxiliary input. To synthesize waveforms from SCENT model outputs, a WaveNet vocoder conditioned on Mel-spectrograms is built. This may be an excellent point to bring up. The proposed solution can produce enough duration conversion, which is a complex problem in conventional techniques. The experiment results show that the proposed approach, using Gaussian mixture models and deep neural networks as acoustic models, outperformed the baseline techniques using Gaussian mixture models and deep neural networks as acoustic models in both objective and subjective assessment. Their previous effort earned first place in the Voice Conversion Challenge 2018, but their most recent attempt exceeded that achievement. The efficacy of different components of this strategy was shown via ablation studies [34].

In another research conducted by Zhang et al., they used three distributed SGD algorithms: Synchronous (SYNC), Asynchronous Decentralized Parallel SGD (ADPSGD), and the hybrid these two algorithms: HYBRID. The acoustic model employed is an LSTM with six bi-directional layers, each of which comprises 1,024 cells (512 cells in each direction). A linear projection layer with 256 hidden units on top of the LSTM layers is followed by a softmax output layer with 32,000 units corresponding to context-dependent HMM states. The LSTM is trained using feature subsequence of length 21, after which it is unrolled to produce LSTMs with different numbers of frames. FMLLR (40-dim), iVector (100-dim), and logmel (40-dim 3) feature input is a mixture of those functions. A training data set from Broadcast News, Conversation, Switchboard, Fisher, and Gigaword was made accessible to the language model (LM) reconstruction algorithm. ADPSGD delivers the best speed-up, whereas HYBRID does not beat SYNC-DDL. They achieved 13.1 per cent WER on the CallHome (CH) dataset and 7.6 per cent WER on the Hub5-2000 Switchboard (SWB) dataset [36].

Tjandra et al., in a study, found out that due to the simplified pipeline compared to the conventional method and its good performance, end-to-end sequence modelling has become a preferable option for automated speech recognition. Nonetheless, end-to-end ASR model training is susceptible to numerous problems, in which the ground truth target transcription and speech features are made contingent on the time-step prediction on the source site. Since the model does not have access to the target sequence ground truth, the inference round is different. It follows that if errors are made, the decoding output gets worse and worse with time. Training and assessment goals often conflict, and this problem arises consequently. Maximum likelihood estimation criteria are used as the objective function during the training phase. The Levenshtein distance and the word mistake rate, on the other hand, are used to assess the quality of ASR systems. In conclusion, they present a technique for optimizing the end-to-end speech recognition model using policy gradient, a reinforcement learning method. A reward function is used to optimise the model coupled to the ASR evaluation criteria (e.g., negative Levenshtein distance). According to the approach suggested, the study's findings show that the model's overall performance is much improved than using just the objective function with the most significant probability and instructor forcing [32].

Pre-trained audio neural networks (PANNs) based on the substantial Audio Set dataset were suggested in a study conducted by Kong et al. They may be used in any other audio-related activity, including those connected to post-production. They presented Wavegram-Logmel-CNN, a different architecture based on log-mel spectrogram and waveform characteristics. To enable the usage of information in both log-mel spectrograms and time-domain waveforms, they merged the log-mel spectrogram and Wavegram to create a new representation. This combination is often used along with the channel dimension. The Wavegram compliments the log-mel spectrogram by offering complementary information. To build PANNs, a wide variety of neural networks were investigated. They also investigate the computational complexity of PANNs. Finally, they transferred PANNs to six different audio pattern recognition works (ESC-50, DCASE_2019 Task1, DCASE_2018 Task2, MSoS, GTZAN, RAVDESS), and exhibit cutting-edge performance in a few of those tasks. They can obtain a state-of-the-art mean average precision (mAP) of 0.439 in their best PANN system with AudioSet tagging, surpassing their previous best system with a mAP of 0.392. PANNs may be helpful when fine-tuned on a small amount of data for new job functions [19].

Human specialists modified the front-end speech augmentation settings for each environment and acoustic model to maximize automated speech recognition accuracy. This research by Kawase et al. has created a unique approach for automatically changing the front-end speech augmentation to maximize the accuracy of a particular ASR engine in this work. When ambient noise fluctuates, the suggested technique allows users to use ASR through electronic devices with less stress. By grouping the environments based on noise characteristics, a genetic algorithm (GA) creates parameter values for front-end speech augmentation for specific environments, which may be dynamically allocated to input speech signals [18].

End-to-end ASR architecture that transcribes voice to text even with no pre-trained alignments has recently made significant progress. The hybrid CTC/attention based Automatic Speech Recognition architecture, which takes advantage of both attention and CTC, is a popular end-to-end technique. The hybrid end-to-end automatic speech recognition architecture exhibit performance comparable to conventional DNN/ hidden Markov model (HMM) speech recognition approaches. But these systems are hard to deploy for online speech recognition. Therefore, Miao et al. In research proposed an online, hybrid connectionist temporal classification (CTC) and attention end-to-end ASR architecture. All offline components of a traditional CTC/attention ASR design are replaced with their streaming counterparts.

- They introduced stable monotonic chunk-wise attention (sMoChA) to replace traditional global attention and monotonic truncated attention (MTA)
- A truncated CTC prefix score is proposed to speed up the calculation of CTC prefix scores.
- Dynamic waiting joint decoding (DWJD) technique is created for dynamically gather CTC and attention predictions in real-time.
- latency-controlled bidirectional long short-term memory (LC-BLSTM) is used to stream the commonly used offline bidirectional encoder network.

Experiments done with HKUST Mandarin and LibriSpeech English tasks show that their suggested online CTC/attention model enhances the real-time factor in human-computer interaction services and maintains performance with low degradation compared to the offline CTC/attention model. This is the first study to give a full-stack real-time/online solution for end-to-end automatic speech recognition architecture for CTC/attention [23].

Far-field automatic speech recognition (ASR), machine recognition of speech given at a distance from microphones, has gotten much attention in science and industry, thanks to a significant boost in speech recognition accuracy. Moreover, it has joined the consumer market with an intelligent personal assistant (IPA) or digital home assistants, the most notable application of a spoken language interface. However, due to the multiple acoustic distortions that arise due to speech recorded at a distance, unique processing pipelines have developed compared to close-talk speech recognition. This paper given by Haeb-Umbach et al. stresses the need for multi-channel speech augmentation for far-field ASR and gives a detailed explanation of current enhancement approaches. Furthermore, when DL was combined with robust signal processing, the performance was considerably improved compared to prior signal processing-only methods [13].

Chai et al. proposed a Cross-Entropy-Guided Measure (CEGM) for evaluating the performance of Speech Recognition. ASR is indirectly assessed by using a voice enhancement front-end, and without performing ASR trials, a novel metric (CEGM) to evaluate the cross-entropy-guided ASR performance of degraded speech is suggested.

The proposed CEGM is calculated in three steps:

- low-level representations using feature extraction,
- high-level non-linear mapping using an acoustic model, and
- a final CEGM between clean and enhanced speech high-level representations.

An illustration of this concept is a priori-generated hybrid acoustic model of the target ASR system built up from a conventional hybrid acoustic model of the source ASR system, which is built up from source ASR system components. The computation of the CEGM then follows this process by applying a cross-entropy criterion. The CEGM model can be used as an objective function for DNN-based voice augmentation instead of using the conventional MMSE criteria. The front-end augmentation model is, thus, able to be modified to enhance the back-end automatic speech recognition system's accuracy. Evaluating the quality of ASR using front-end functions of the voice enhancement algorithm (FI) does not directly assess single-channel ASR functionality. This implies processing speech with unprocessed noisy speech and MMSE-optimized enhancement for ASR systems employing fixed multi-condition acoustic models is better than processing speech with MMSE-optimized enhancement in deep architectures. A CEGM CHiME-4 challenge show was done to conduct a WER (costly to calculate) and ASR (an objective measure of ASR performance) evaluation, and results in correlating with WER and in the most accurate assessment of ASR performance were successful. In the WER testing, speech enhancement used for automated speech recognition systems in which the multi-condition acoustic models are fixed may have been less effective for CEGM-optimized speech enhancement than unprocessed speech or improved speech generated using MMSE-optimized enhancement [7].

The research by Chuang et al. focuses on improving ASR or end-to-end Speech Translation models in general. It uses word embeddings, which are continuous representations of words, to improve automatic speech recognition in cascaded systems and end-to-end speech translation models. The advantage of utilizing word embedding is that it is simple to construct word embedding by training on pure textual data, which solves data scarcity. Word embedding also offers speech models with additional contextual information. They were inspired to turn word embedding information into speech models. They suggest a new decoding approach to fuse the semantic links among words for further improvement in ASR, and they use word embeddings as a regularize to lower the Word Error Rate. They proposed using word embeddings as an intermediate representation to improve translation performance in the end-to-end Speech translation paradigm. Their findings suggest that voice signals may be mapped to semantic space, which encourages further research into using the proposed approaches in spoken language processing tasks [9].

To achieve reliable end-to-end automatic speech recognition (ASR), the integrated training framework for speech enhancement and recognition approaches has achieved reasonably good results. On the other hand, these methods use the augmented feature as an input to the voice recognition component, influenced by the speech distortion issue. The research by Fan et al. offers a gated recurrent fusion (GRF) technique with a combined training architecture for robust end-to-end ASR to overcome this problem. The noisy and improved features are dynamically combined using the GRF method. As a result, the GRF not only can eliminate noise signals from improved features, but it can also learn the basic delicate structures from noisy features to reduce speech distortion. Speech augmentation, GRF, and speech recognition are all part of the suggested method. The input speech is enhanced first by using a mask-based speech enhancement network.

Second, the GRF is utilized to deal with voice distortion. Third, the speech transformer algorithm is used as the speech recognition feature to increase ASR performance. Finally, the hybrid training framework is used to optimize these three components simultaneously. The proposed method reduces the relative character error rate (CER) by 10.04 per cent compared to the standard joint enhancement and transformer method that solely uses enhanced features, according to experimental results. Especially for the low signal-to-noise ratio (SNR), this method achieved better performances with a 12.67% CER reduction [10].

LM-based Certainty Approach suggests that a considerable amount of work is required for automatic transcribe recognition from speech data. Getting better models for a cheaper cost should include choosing samples with more information from the uncertified data pool. ASR's active learning has therefore been deemed a study field. Two new active learning methods, such as the SMCA and the LM-based Certainty Approach, were used in this study. These two techniques look at data from two different points of view, and they may be used together to take advantage of collaborative sampling in certain circumstances. A state-of-the-art committee-based approach known as heterogeneous neural networks was used in these studies (HNN). This CER method was used in a Chinese Mandarin voice recognition problem, and the findings indicate that it is simple to apply and provides the most significant results [31].

8.6 Conclusion and Future Scope

Speech research has been the first step toward regular man-machine communication throughout the last few decades. This paper aims to provide a complete overview of speech recognition research and a year-by-year analysis of development and status. Although tremendous progress has been made in recent decades, there is still further work to be done. Currently, research is focused on establishing and constructing systems that are considerably more resilient to changes in the acoustic environment, speaker characteristics, language characteristics, external noise sources, and other factors.

References

1. M. Abdel-Mottaleb, N. Dimitrova, R. Desai, J. Martino, Conivas: content-based image and video access system. In *Proceedings of the Fourth ACM International Conference on Multimedia*, MULTIMEDIA '96, pp. 427–428, New York, NY, USA, 1997. Association for Computing Machinery
2. J. Adcock, M. Cooper, L. Denoue, H. Pirsiavash, L.A. Rowe, Talkminer: a lecture webcast search engine. In *Proceedings of the 18th ACM International Conference on Multimedia*, MM '10 (New York, NY, USA 2010), pp. 241–250. Association for Computing Machinery
3. T. Afouras, J.S. Chung, A. Senior, O. Vinyals, A. Zisserman. Deep audio-visual speech recognition. *IEEE Transactions on Pattern Analysis and Machine Intelligence*, p. 1 (2018)
4. M.A. Anusuya, S.K Katti, Speech recognition by machine, a review. *arXiv preprint* (2010). arXiv:1001.2267
5. S.J. Arora, R.P. Singh, Automatic speech recognition: a review. Int. J. Comput. Appl. **60**(9) (2012)
6. A. Biswas, A. Gandhi, O. Deshmukh, Mmtoc: a multimodal method for table of content creation in educational videos. In *Proceedings of the 23rd ACM International Conference on Multimedia*, MM '15 (New York, NY, USA, 2015), pp. 621–630. Association for Computing Machinery
7. Li. Chai, Du. Jun, Qing-Feng. Liu, Chin-Hui. Lee, A cross-entropy-guided measure (cegm) for assessing speech recognition performance and optimizing dnn-based speech enhancement. IEEE/ACM Trans. Audio, Speech Lang. Proc. **29**, 106–117 (2021)
8. C.-C. Chiu, T.N. Sainath, Y. Wu, R. Prabhavalkar, P. Nguyen, Z. Chen, A. Kannan, R.J. Weiss, K. Rao, E. Gonina, N. Jaitly, B. Li, J. Chorowski, M. Bacchiani, State-of-the-art speech recognition with sequence-to-sequence models. In *2018 IEEE International Conference on Acoustics, Speech and Signal Processing (ICASSP)*, pp. 4774–4778 (2018)
9. Shun-Po. Chuang, Alexander H. Liu, Tzu-Wei. Sung, Hung-yi Lee, Improving automatic speech recognition and speech translation via word embedding prediction. IEEE/ACM Trans. Audio, Speech Lang. Proc. **29**, 93–105 (2021)

10. Cunhang Fan, Jiangyan Yi, Jianhua Tao, Zhengkun Tian, Bin Liu, Zhengqi Wen, Gated recurrent fusion with joint training framework for robust end-to-end speech recognition. IEEE/ACM Trans. Audio, Speech Lang. Proc. **29**, 198–209 (2021)
11. Gregory Gelly, Jean-Luc. Gauvain, Optimization of rnn-based speech activity detection. IEEE/ACM Trans. Audio, Speech Lang. Proc. **26**(3), 646–656 (2018)
12. Hossein Hadian, Hossein Sameti, Daniel Povey, Sanjeev Khudanpur, End-to-end speech recognition using lattice-free mmi. Proc. Interspeech **2018**, 12–16 (2018)
13. Reinhold Haeb-Umbach, Jahn Heymann, Lukas Drude, Shinji Watanabe, Marc Delcroix, Tomohiro Nakatani, Far-field automatic speech recognition. Proceedings of the IEEE **109**(2), 124–148 (2021)
14. C. Hui, S. Yunyu, Y. Haisheng, G. Ming, Yongxiang Liu Xiang, Xia, A fast and robust key frame extraction method for video copyright protection. *J. Elect. Comp. Engin.* (March 2017)
15. S. Jothilakshmi, Spoken keyword detection using autoassociative neural networks. Int. J. Speech Technol. **17** (2014)
16. C.H. Lee, B.H. Juang, W. Chou, Statistical and discriminative methods for speech recognition. *The Kluwer International Series in Engineering and Computer Science (VLSI, Computer Architecture and Digital Signal Processing)* (1996)
17. V.K. Kamabathula, S. Iyer, Automated tagging to enable fine-grained browsing of lecture videos. In *2011 IEEE International Conference on Technology for Education*, pp. 96–102 (2011)
18. Tomoko Kawase, Manabu Okamoto, Takaaki Fukutomi, Yamato Takahashi, Speech enhancement parameter adjustment to maximize accuracy of automatic speech recognition. IEEE Trans. Consum. Electr. **66**(2), 125–133 (2020)
19. Qiuqiang Kong, Yin Cao, Turab Iqbal, Yuxuan Wang, Wenwu Wang, Mark D. Plumbley, Panns: Large-scale pretrained audio neural networks for audio pattern recognition. IEEE/ACM Trans. Audio, Speech Lang. Proc. **28**, 2880–2894 (2020)
20. M. Lin, J.F. Nunamaker, M. Chau, H. Chen, Segmentation of lecture videos based on text: a method combining multiple linguistic features. In *37th Annual Hawaii International Conference on System Sciences, 2004. Proceedings of the*, p. 9 (2004)
21. M. Mahrishi, S. Morwal, Index point detection and semantic indexing of videos a comparative review. *Advances in Intelligent Systems and Computing AISC Springer* (2020)
22. M. Merler, J.R. Kender, Semantic keyword extraction via adaptive text binarization of unstructured unsourced video. In *2009 16th IEEE International Conference on Image Processing (ICIP)*, pp. 261–264 (2009)
23. Haoran Miao, Gaofeng Cheng, Pengyuan Zhang, Yonghong Yan, Online hybrid ctc/attention end-to-end automatic speech recognition architecture. IEEE/ACM Trans. Audio, Speech Lang. Proc. **28**, 1452–1465 (2020)
24. J. Pustejovsky, A. Stubbs, Natural language annotation for machine learning
25. R. Rana, R. Singh, D. Mishra, An improved hindi speech recognition system by using i-rover (2013)
26. Mirco Ravanelli, Philemon Brakel, Maurizio Omologo, Yoshua Bengio, Light gated recurrent units for speech recognition. IEEE Trans. Emerging Topics Comput. Intell. **2**(2), 92–102 (2018)
27. M. Riedl, C. Biemann, TopicTiling: a text segmentation algorithm based on LDA. In *Proceedings of ACL 2012 Student Research Workshop* (Jeju Island, Korea, July 2012), pp. 37–42. Association for Computational Linguistics
28. Florinda Sauli, Alberto Cattaneo, Hans van der Meij, Hypervideo for educational purposes: a literature review on a multifaceted technological tool. Technol. Pedag. Educ. **27**(1), 115–134 (2018)
29. M. Sharma, K. Sarma, *Soft-Computational Techniques and Spectro-Temporal Features for Telephonic Speech Recognition: An Overview and Review of Current State of the Art*, 11 (2015)
30. R. Sharma, M. Mahrishi, S. Morwal, G. Sharma, Index point detection for text summarization using cosine similarity in educational videos. IOP Conf. Series Mater. Sci. Eng. **1131**(1), 012001 (Apr 2021)

31. Xiusong Sun, Bo. Wang, Shaohan Liu, Lu. Tingxiang, Xin Shan, Qun Yang, Lmc-smca: A new active learning method in asr. IEEE Access **9**, 37011–37021 (2021)
32. Andros Tjandra, Sakriani Sakti, Satoshi Nakamura, End-to-end speech recognition sequence training with reinforcement learning. IEEE Access **7**, 79758–79769 (2019)
33. N.J. Uke, R. Thool, Segmentation and organization of lecture video based on visual contents. *Int. J. e-Education, e-Business, e-Management and e-Learning* (2012)
34. Jing-Xuan. Zhang, Zhen-Hua. Ling, Li-Juan. Liu, Yuan Jiang, Li-Rong. Dai, Sequence-to-sequence acoustic modeling for voice conversion. IEEE/ACM Trans. Audio, Speech Lang. Proc. **27**(3), 631–644 (2019)
35. Lin Zhang, Lu. Yao, Video object segmentation by latent outcome regression. IEEE Access **8**, 30355–30367 (2020)
36. W. Zhang, X. Cui, U. Finkler, B. Kingsbury, G. Saon, D. Kung, M. Picheny. Distributed deep learning strategies for automatic speech recognition. In *ICASSP 2019–2019 IEEE International Conference on Acoustics, Speech and Signal Processing (ICASSP)*, pp. 5706–5710 (2019)
37. Tianxiang Zhou, Ke Wang, Jun Wu, and Ruifeng Li. Video text processing method based on image stitching. In *2019 IEEE 4th International Conference on Image, Vision and Computing (ICIVC)*, pp. 561–566 (2019)

Chapter 9
Anomaly Detection in Industrial IoT Applications Using Deep Learning Approach

Chetan Bulla and Mahantesh N. Birje

Abstract Internet of Things (IoT) becomes popular in last two decade as it provides many advantages such as flexibility, autonomous, cost effective and productivity. Many industries adopted IoT to improve the efficiency, security and predictive maintenance. To improve the quality of service (QoS), it is essential to identify various types of anomalies in productive maintenance. An anomaly is a value, a status of resources or outcome that deviates from expected or normal values and it affects QoS of production. In this paper, a multi-agent based anomaly detection scheme is introduced to improve the QoS. The fog computing infrastructure is used to reduce the latency of communication. Multiple agents are deployed in fog node to perform the various operations of detecting anomalies. The proposed anomaly detection scheme uses a multi-step prediction technique and applies an anomaly detection algorithm to detect anomalies. The Gated Recurrent Unit (GRU) model is used for Multi-step time series prediction and a bio-inspired Artificial Bee Colony algorithm is used for tuning the GRU model hyperparameters to improve the accuracy. The proposed model detects various types of anomalies in the fog computing environment. The Google Colab using TensorFlow library Keras is used for experimental evaluation. The proposed model increases accuracy over existing approaches, according to the experiment evaluation.

Keywords Multi-step prediction · Anomaly detection · Hyperparameter optimization · Prediction · GRU · Cloud computing · And fog computing

C. Bulla (✉)
Department of CSE, KLECET, Chikodi, India

M. N. Birje
Department of CSE, VTU, Belagavi, India

S. Fernandes et al. (eds.), *Artificial Intelligence in Industrial Applications*, Learning and Analytics in Intelligent Systems 25,
https://doi.org/10.1007/978-3-030-85383-9_9

9.1 Introduction

Internet of Things (IoT) is the new emerging technology that interconnects various types of computing devices, mechanical devices, and digital machines [1]. Each of these devices contain different types of sensors that generate and transfer data over the cloud for processing without human-to-computer interaction. IoT was initially introduced for industry and business applications, but now it is used in offices, homes, and hospitals. The main reason behind this change is the usage of intelligent devices in everyday life. These intelligent devices are cheaper, easy to use, low maintenance costs, and can work in any environment. Time-sensitive IoT applications require a shorter response time and mobility support, such as smart hospitals and smart cities. Cisco has introduced the fog computing where computational resources are brought to the network edge to resolve the bandwidth and latency issue [2, 4].

The monitoring system [6] providing accurate information of fog computing resources ad associated IoT devices. It keeps track of all data flow and resources' status [8] and alerts the users for abnormal conditions. To differentiate between normal and abnormal systems, it is essential to develop an anomaly detection model. Anomaly detection is one of the major concerns in improving performance and troubleshooting activities in fog computing to improve QoS [5]. An anomaly is a value, a status of resources or outcome that deviates from expected or normal values/status. It is also helpful to identify the malfunction and anticipate failure. The major reasons for anomalies are resource competition, resource bottleneck, misconfigurations, software defects, hardware failures, and external attacks in fog computing infrastructure. Early detection of these anomalies improves the reliability and availability and reduces the shutdown time and maintenance cost.

Different types of anomaly can be found at different layers of fog computing infrastructure. Figure 9.1 shows the fog computing architecture with possible anomalies [7]. There are three layers: Cloud layer, Fog layer, and Terminal layer. The terminal layer contains various sensors and actuators that sense the environment and transfer the data to the fog layer. The fog layer is an intermediate layer and contains limited computational capacity such as Gateways and fog servers. The fog layer checks the processing requirements and processes the data. If the fog layer is unable to process then it is sent to the cloud layer. The cloud and fog computing infrastructure suffer from CPU hog, memory leak, and Workload contentions. The CPU hog means the processor consumes all its processing cycles by injecting infinite loops or viruses. Memory leaks suffer from memory exhausts by injecting anomalies to consume all memory capacity, and workload contention leads to saturation in resource components by emulating user requests.

Most fog-based applications use IoT devices or sensors that are highly dependent on accurate operation, which is reasonably difficult to guarantee. These sensors are inexpensive and prone to malfunction or error. In addition, sensors are installed under potentially harsh weather conditions, making their breakage much more probable [9]. The reasons for the sensor's anomalies are Spike, Drift, and constant. The Spike is a short duration in sensed values that diverges from other measurements; Drift is

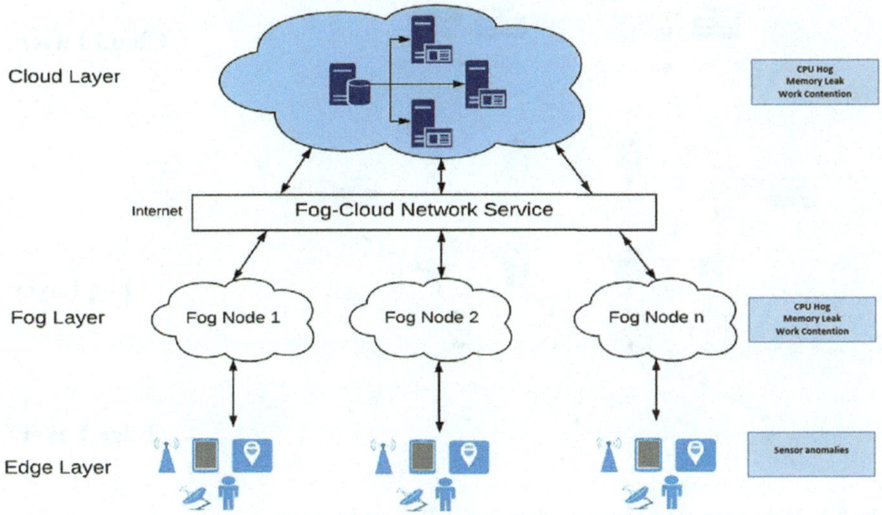

Fig. 9.1 Fog computing architecture with possible anomalies

off-set in the measurements of sensor values and a sensor reports as a constant value in a different scenario. The high probability of erroneous readings or data corruption occurs during data transmission between sensors and fog nodes. The large variations in sensed values and read data in fog nodes is called a sensor anomaly. The anomalies in fog computing increase shutdown time, maintenance cost, energy consumption, computational overhead and reduce performance, availability, and reliability. So it is essential to detect the anomalies to improve the performance of fog computing services. Anomalies can be classified into the following three categories:

Point anomaly are separate data points that are considered anomalous with respect to the remaining data. This type of anomaly refers to data points that are given without contextual aspects, such as time. *Collective* anomaly refers to a set of continuous data points, which are anomalous w.r.t the complete dataset. Two notes are worth mentioning: firstly, individual data points in such an anomalous collection may not be anomalies by themselves, while only their common existence shows up as an anomaly. Secondly is that this type of anomaly can appear in both context-aware data and contextless data. *Contextual anomalies* are individual data points or collection of data points are considered as anomalous with respect to particular context or applications.

In this paper, the multiple agents system (MAS) is used to perform the various operations in detecting anomalies [6]. These agents are installed in each fog node to collect the data, prepare the data, anomaly detection, and hyperparameters tuning. The agents are intelligent programs, and these programs work cooperatively and collaboratively to perform a variety of tasks. Figure 9.2 shows the architecture of a multi-agent system in a fog computing environment. The architecture is divided into 3 layers: cloud, fog and IoT layer. Cloud layer contains large datacenters. The fog

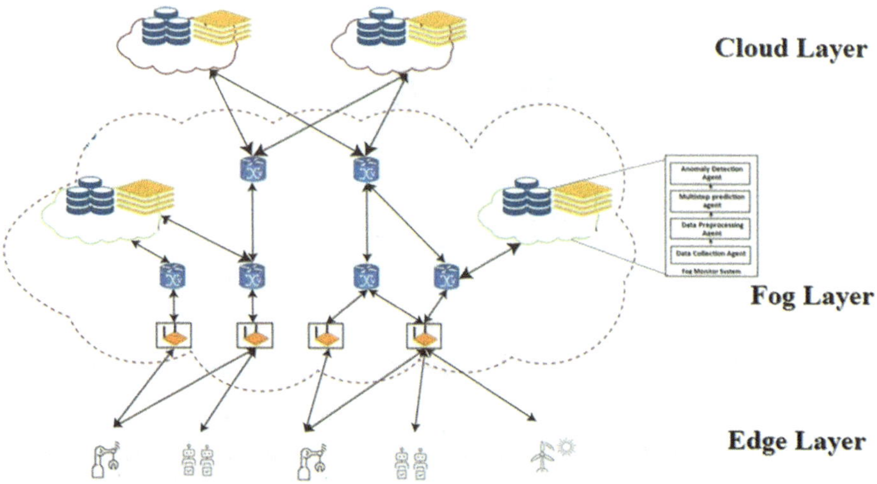

Fig. 9.2 Multi-agent based anomaly detection in IIoT environment

node have limited computational resources. The IoT devices contains various smart devices and its sensors. The traditional sugar cane factory slowly converting to smart industries by adapting automation in manufacturing process. The various sensors are deployed in smart machines/devices to check proper operations. The smart devices are connection to fog computing infrastructure for computational services.

Various IoT applications such as Industry IoT, Medical IoT, and smart city, etc. are used in the fog computing environment. These applications install multiple sensors and actuators. The data generated by the fog node, virtual machine, and sensors are gathered by the collection agent [3] and sends them to the data preparation agent. The data preparation agent removes outliers, noise, and normalize the data. The anomaly detection agent predicts future time series and identifies anomalies. Hyperparameter tuning for GRU model using artificial bee colony is carried out by hyperparameter optimization agent. The anomaly detection agent compares the predicted value and actual values of data. If the difference between these is more than a certain threshold, then it is identified as an anomaly. The main contributions of this proposed work are: multi-step prediction on time series data is designed using GRU Model. An Artificial Bee Colony based hyperparameter optimization algorithm is implemented to increase the accuracy of multi-step prediction. An adaptive threshold mechanism is used to identify anomaly in Industrial IoT.

The remaining sections are organized as follows: Sect. 9.2 discusses state of art work on anomaly detection methods. Section 9.3 presents anomaly detection model for fog computing environment. Section 9.4 discusses the performance evaluation of the proposed anomaly detection model. Finally paper is summarized and concluded in Sect. 9.5.

9.2 Related Work

In real-time applications, anomaly detection and prediction are of great importance in preventing anomaly or reducing damage. An adaptive anomaly detection scheme for IoT data is proposed for Hierarchical Edge Computing [4]. The author proposed three anomaly detection models to match different complexities and associate them with the HEC layer in a bottom-up approach. Hyper ellipsoidal Clustering [11] proposed fog empowered distributed computing. The fog node and cloud node divide the data based clustered and adopt an anomaly detection algorithm. The proposed solution compared two traditional anomaly detection models, (i) Centralized and (ii) Distributed algorithm. The results show a specific improvement in the accuracy of the anomaly detection model.

A lightweight anomaly detection model for Internet of Things proposed in [15]. The proposed model is host based anomaly detection model uses two popular clustering technique inverse Weight Clustering (IWC), and decision tree. The model initially divide the dataset into clusters and apply decision tree for classification. To evaluate the model, the authors used only two sensors, temperature, and humidity. IWC is a modified version of k-means, run unlimited iterations to get optimal solutions. To detect and predict anomalies at runtime to optimize system performance, Hidden Markov Model-based anomaly detection and the prediction model is developed [10]. The proposed model considered the problem of mapping observations and identified anomalous behavior edge cluster workloads. The authors extracted variations in the containers' workload over various timestamps captured by Hidden Markov Models (HMM) and it is used to calculate model parameters from container workload data.

The different machine learning models [14] for anomaly detection in IoT has been proposed. To evaluate and compare these machine learning models accuracy, and receiver operating characteristic curve are used. The ANN, Random Forest, and Decision Tree models recorded the accuracy up to 98.6%. To detect various threats and false alarms, a machine learning-based anomaly detection model is developed. A non-parametric Online Locally Weighted Projection Regression (OLWPR) is used to detect wireless sensors networks anomalies. Various parameters like recall, precision, accuracy, RMSE, and MAPE are used to evaluate the proposed model. The experiment evaluation shows that the anomaly detection model improves accuracy compared to the Dynamic Bayesian Network (DBN).

The most advanced and dynamic reinforcement learning algorithm called asynchronous advantage actor-critic (A3C) algorithm addresses the anomaly detection problem. The experiment results show improvement in recall and precision compared to other machine learning based anomaly detection models. A vector convolutional deep learning (VDCL) model is used to identify outliers in IoT. The main objective is to improve the scalability of anomaly detection in the non-scalable nature of network traffic records in IoT. The master fog node uses the VCDL model to classify the known and unknown anomalies. The experimental results showed that the proposed framework improved accuracy, precision, and recall compared with the existing anomaly detection system. An anomaly detection method uses multivariate data using a convolution neural network [15] to improve accuracy. The proposed model behaves human vision and decision making to detect anomalies. CNN is trained to identify and detect anomalies in multivariate time series data. During training representation and description of anomalies were learned. Experiment results show that proposed anomaly detection is scalable, accurate, and faster.

9.3 Proposed Model

The various IoT devices from different locations are connected to the fog nodes. The IoT devices generate large amount of data repeatedly. These raw data is collected and apply filtering to convert the data in a appropriate format for training. The proposed anomaly detection model works in three stages (i) Data preprocessing (ii) Multistep prediction and (iii) Anomaly detection. The data preprocessing is used to prepare the data for training and testing purposes. The proposed model utilizes the fog monitoring system to collect, filter, and normalize the data coming from IoT devices. The GRU model with hyperparameter tuning is used for multistep prediction. (iii) Anomaly detection compared the actual data with predicted data and consider certain points as anomaly if the difference between actual and predicted data is more than threshold value. Figure 9.3 shows the proposed anomaly detection model. The following sub sections discusses the working of these stages.

1. **Data Preprocessing**

The quality of data is very crucial in training the deep learning. The data preprocessing is the process of removing unwanted data and making it suitable for deep learning models. The proposed system utilizes the fog monitoring system to collect, filter, and normalize the data coming from IoT devices. The monitoring system continuously keeps track of all the information of fog computing infrastructure. It monitors all resources' health, Event data, and data generated by IoT devices in fog computing.

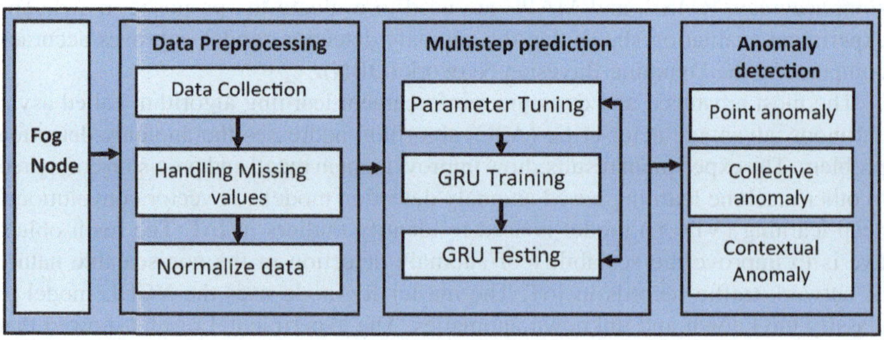

Fig. 9.3 Anomaly detection with hyperparameter tuning model

The proposed model uses a fog monitoring system [3] for data preprocessing tasks such as data collection, aggregation, data filtering, handling missing values, and data normalization. The quality of data is one of the key factors for any deep learning model to improve the performance. In the proposed model, two data processing strategy (i) handling missing data and (ii) Normalize the data are adopted. To handle the missing data, the mean, median, or mode of a particular feature is calculated and added to missing values. In order to avoid collecting duplicate data frequently, the model collects the temporal data if there is a certain difference between previous data and current data.

$$CD_t = \frac{D(t) - D(t-1)}{MAX} X\,100 \tag{9.1}$$

The mean and median values of a particular feature is used to fill up the missing values in training data.

$$\underline{x} = \frac{\sum_{i=1}^{N} x}{N} \tag{9.2}$$

$$Median = \left(\frac{n+1}{2}\right)^{th} term\ (if\ n\ is\ even) \tag{9.3}$$

$$Median = \frac{\left(\frac{n}{2}\right)^{th} term + \left(\frac{n}{2}+1\right)^{th} term}{2} \text{(If n is odd)} \tag{9.4}$$

Further, it is essential to normalize the outliers in the data. An outlier is a data point that differs significantly from other observations. The outliers in data may reduce the performance of the deep learning model. So it is essential to normalize the data to a certain range (min, max). The outlier is normalized using:

$$X' = \frac{x - min(x)}{max(x) - min(x)} \tag{9.5}$$

These processed data are used for training GRU network prediction model to identify anomalous data.

2. **Multistep-Prediction**

The proposed model identifies anomalies based on multi-step prediction. It compares actual data with predicted data and considers an anomaly if the difference between these two data is greater than a threshold value. The GRU is a type of Recurrent Neural Network (RNN) used to learn temporal data interference like time-series data prediction and find a particular pattern in time series data. GRU is stronger than RNN, as it resolves the problem vanishing gradient, learns from inputs that are independent of each other, and can store longer memories. GRU networks are well suited to analyzing, classifying, and making forecasts based on time series data, as there can be unfamiliar period lags between significant events in time series [11]. So in this paper, the GRU network model is used to for multistep time series prediction. Figure 9.4 shows the GRU unit with its components.

The GRU has two gates: update and reset gate. These two gates regulate the information/data flow inside the GRU cell. An input gate allows only necessary input values in the GRU cell. It decides what information to remove and what new information to be added. The reset gate is used to decide how much past information to forget. Initially update gate z_t for time step t is calculated using following formula.

$$Z_t = \sigma(Wt_f.X_t + U^{(z)}h_{t-1}] \tag{9.6}$$

When *x_t* is plugged into the network unit, it is multiplied by its own weight *W(z)*. The same goes for *h_(t-1)* which holds the information for the previous *t-1* units and is multiplied by its own weight *U(z)*. Both results are added together and a sigmoid activation function is applied to squash the result between 0 and 1. next reset gate is used to decide how much past information to remove the network. The reset gate is can be calculated using

$$r_t = \sigma(Wt_f.X_t + U^{(r)}h_{t-1}] \tag{9.7}$$

the reset gate formula is similar to update gate. The weight and gate's usage are different in both gates. the current memory content is introduced to reset gate to store the relevant information from the past. the current memory content is calculated as follows:

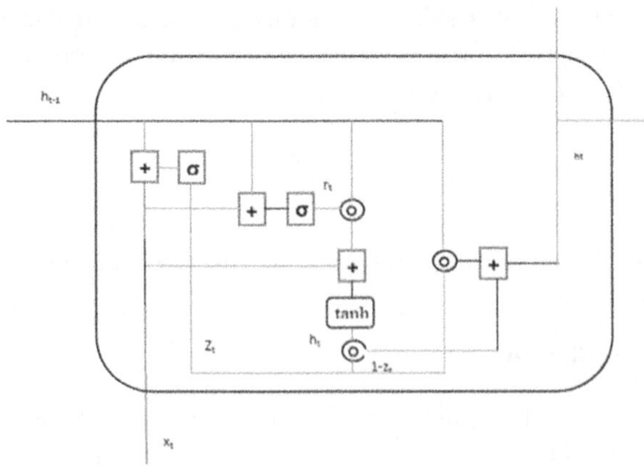

Fig. 9.4 GRU model

$$h_t = tan_h(W_x.O_t + r_i.Uh_{t-1}) \tag{9.8}$$

as the last step, the network needs to calculate h_t vector which holds the information for the current unit and passes it down to the network. it also determines what to collect from the memory content. The following formula identifies the important information in memory content.

$$h_t = Z_t.h_t + (1 - z_t).h_t^{'} \tag{9.9}$$

1. Apply element-wise multiplication to the update gate z_t and h_(t-1).
2. Apply element-wise multiplication to (1-z_t) and h'_t.
3. Sum the results from step 1 and 2.

The model can learn to set the vector z_t close to 1 and keep a majority of the previous information. Since z_t will be close to 1 at this time step, 1-z_t will be close to 0 which will ignore big portion of the current content (in this case the last part of the review which explains the book plot) which is irrelevant for our prediction.

In order to prevent the units from being strongly co-adapted, it is necessary to drop a few secret units along with their relationship during the training process. After dropout, each secret device will become more versatile and will be able to build valuable functionality on its own without relying on other units to help minimize overfitting. The dropout technique is used to reduce overfitting problem and improve the performance of neural networks applications. In the proposed model, dropout is used only in hidden layers. The main idea is to feed the normal data to GRU based Neural Network for training the GRU network to predicts the next 'n' time steps. The predicted values are compared to actual values and considered anomalous if the difference is larger than the specified threshold. Based on the density of anomalous points, it is categorized as warning, error, and issue.

The multi-step time-series prediction model is designed using GRU Model. Given existing time series data, $X = \{x_1, x_2 \ldots x_{t-1}\}$, where x_i denotes values at the time-stamp *i*, the task is to predict the future value x_t. The sliding window model is adopted to collect the recent values for multi-step prediction. For the prediction $\{x_{t-w}, x_{t+1-w} \ldots x_{t-1}\}$ is used, where 'w' is the window size. These sequences are divided into mini-batches and fed into GRU network for training with 'n' epochs.

$$y_{t+n} = O_n(y_t, y_{t+1}, y_{t+2} \ldots .y_{t+n-1}) \tag{9.10}$$

where "n" is the number of steps to forecast into the future, O_n is any GRU Learning model outcome. Figure 9.5 shows a window-based multistep prediction. The main advantage of GRU is it remembers existing values while predicting future values. The window size is 10 and sliding towards the right side 'n' step after predicting last value. The process will continue until specified number of multistep prediction is required.

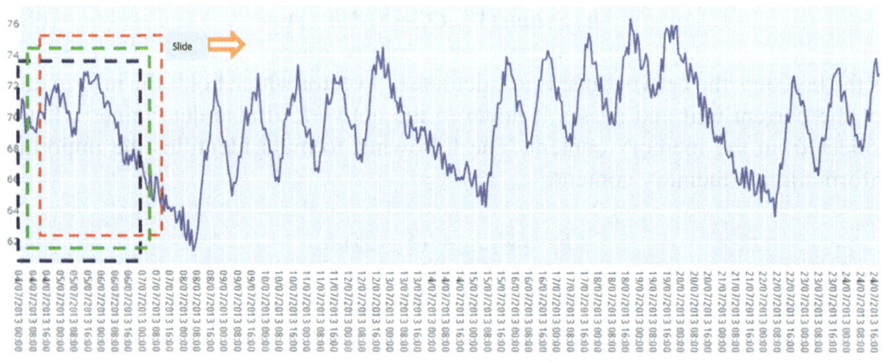

Fig. 9.5 Multistep prediction

2.1. **Hyperparameter optimization using Artificial Bee Colony.**

The ABC is a bio-inspired algorithm mimics bee colonies' intelligent behavior in foraging [13]. It has been used in various application to optimize the performance. The colony of bees in ABC is classified into three types. Employee bee searches the food around of food sources in their memory and shares knowledge to other bees, onlooker bee uses this knowledge and find the good quality of nectar. Scout bees find new food sources and converted themselves to employee bees. The proposed model uses ABC algorithm for hyperparameter tuning to improve the accuracy.

The hyperparameters are used to control the learning process in machine learning applications [15]. The performance of the model can be improved by optimizing hyperparameter values. The main objective of ABC optimization techniques is to choose the best combination values of hyperparameters to improve predictions' accuracy. Automatic Hyperparameter optimization has several advantages: first, it reduces human efforts in deploying machine learning models as a manual assigning hyperparameter to consume more computational resources, and it is a tedious process. The Hyperparameter configurations are dependent on the type of problem and its complexity. Second, the machine learning model improved performance by identifying the most suitable Hyperparameter values for the target application. Further, it improves overall performance with more reproducible results.

In the proposed anomaly detection model, a bio-inspired algorithm called an Artificial Bee Colony (ABC) algorithm is used for optimizing the GRU to increase the accuracy of prediction. The ABC algorithm is used to select appropriate hyperparameters of the GRU model. In ABC Algorithm, solutions to the optimization problems [15]. The Bees are always looking at storing good quality honey for a longer time. The quality of nectar is the main objective feature in processing and storing. The 'n' number of bees are employed for 'n' number of food sources. One onlooker bee is employed for each bee to identify the nectar's quality by taking information from the employed bee. The GRU model adopts ABC foraging behavior to select the best combination of hyperparameters to get the highest accuracy in multistep prediction. The traditional ABC algorithm faces some challenges in exploration and exploitation.

To accelerate the convergence speed of ABC, a population size reduction strategy is introduced in standard ABC algorithm. It will help to increase the explorative search behavior at early stages of the search process. Based a survival strategy and survival rate (SR), it determines the which candidate solutions form the current population survive and pass to the next population. The SR is a constants value and its range is 0 to 1. The following equation represents the population search strategy:

$$NP_{current} = \begin{cases} Max((NP_{current} - NP_{min}), NP_{threshold} & ifSR > r \\ NP_{current} & otherwise \end{cases} \tag{9.11}$$

Further, a modified search strategy is used to improve exploitation behavior of ABC. A new search operator uses information from the food sources of the present population by selecting random pair of information and also best food source. The following equation represents the new search operator

$$v_{ij} = x_{ij} + sf * (x_{best,j} - x_{ij} + x_{r1,j} - x_{r2,j}) \tag{9.12}$$

The hyperparameter in proposed work are: (i) sliding-window-size (oh1), GRU-units (oh2), dropout-rate (oh3), regularizer (oh4), regularizer-rate (oh5), optimizers (oh6), epochs (oh7) and learning rate (oh8). Each hyperparameter has different types and ranges of values. The solutions are obtained in the form of the best amalgamation of hyperparameters that gives the highest accuracy for multi-step prediction problem. These hyperparameter combinations are used to train the GRU network for multistep predictions. Further, the prediction model are evaluated using RMSE and it is considered as the fitness value in ABC algorithm. The ABC algorithm initially assigns the values to the hyperparameter. The GRU model trained with these hyperparameters and calculate the fitness value (RMSE). At the next step, new random hyperparameters values are assigned and check fitness value. The algorithm works iteratively until maximum fitness value is found or maximum cycle numbers (iterations) completed. The new fitness value is recorded if it is better than previously computed fitness value.

3. **Anomaly Detection Model**

The multi-step prediction is used to identify the anomalous data in the temporal data. The observed data are compared with predicted values. If the error rate between the observed value and the predicted value is greater than a predefined threshold value, then it is considered an anomaly. The threshold value varies depending upon the type of application. The following equation is used to calculate the error:

$$E_t = |y_t - x_t| \tag{9.13}$$

where x_t is the timestamp data, $X = (x_1, \times 2, \ldots. x_m)$, y_t is the value of timestamp t in the time series data $Y = (y_1, y_2, \ldots. y_m)$, and E_t is error value. These prediction error are put together in a vector form. The threshold value *Tp* is defined to find the anomalies in temporal data using following equation. The Error rate is higher than threshold T_p then it is considered as anomalous value(s):

$$A = E_t > T_p \tag{9.14}$$

Based on number of anomalies identified, it is classified 'warning',' error' and 'issue'. The proposed model classifies the anomalies as warning, error, or issue depending upon the number of anomalous points identified in the temporal data. The single anomaly point is considered as 'warning' and two anomaly points are considered as 'error'. More than two anomalies are considered as 'issue'. The overall anomaly detection process is shown in Fig. 9.6. The anomaly score is used to identify significance of problem and it is defined as:

$$S_x = \sqrt{(\tilde{x} - x)^T \sum^{-1} (\tilde{x} - x)}, \tag{9.15}$$

where, where Σ is the 'covariance matrix' of the training input, x is the input values and ~x is predicted value. The spike and drift are the common anomalies in time series dataset. To identify spike and drift the following equation are used. The prediction error is used to identify the spike and drift anomalies. The prediction error is more than threshold and it has highest variance then it is considered as *spike*. The prediction error range have the same pattern with high prediction is considered as *drift*. The spike and drift are calculated using following equations

$$Spike_t = E_t > T_p ANDE_t > Var_{max} \tag{9.16}$$

Here, $\mathrm{Var}_{\max}$ is maximum variation, E_t is error rate at time t and T_p is threshold value.

$$Drift_t = \left[Avg(W_{i-1}) - Avg(W_1)\right] > n \tag{9.17}$$

Here, the window values are used to calculate the drift values. The difference between previous average values of window and current average values of windows is greater than *'n'* is considered as drift. Depending upon the types of data and its application, the value of n is set.

9.4 Performance Evaluation

This section discusses experiment setup, performance evaluation of the proposed anomaly detection model. The proposed model is compared with start of art anomaly detection models.

Algorithm: Anomaly detection with hyperparameter optimization using Artificial Bee Colony

1: Initialize the Hyperparameter value (min, max) values, window size, time-steps, Employee Bees (EB), Onlooker Bees(OB), Maximum Cycle Number MCN), Solutions (S_N), k=0, and set the threshold value (0.5)

2. Create Initial population HP_x={SLW, N, DR, LR, RR, OP} and assign predefined standard value

$$x_{ij} = predifined\_value$$

3. Read input values to sliding windows
4. Initialize the counter= 0
5. ***do***

    Counter++
    Assign sliding window values to the GRU network
    Train the GRU network
    Evaluate the fitness value of training using

    $$fitness_i = \{\frac{1}{(1+f_i)}\ if\ f_i \geq 0\ 1 + abs(f_i) if\ f_i < 0$$

    $$p_i = \frac{fitness_i}{\sum_i^{SN} fitness_i} \quad \text{for i = 1.. EB}$$

    Check fitness values the previous solution
        if (p_i<threshold) then
            break
        else
            create a new solution by modifying an old solution (hyperparameters)

    $$v_{ij} = x_{ij} + \emptyset_{ij}\,(x_{ij} - x_{kj})$$

    While(counter >=MCN)

6. Memorize the best hyperparameter combination.
7. Get multi-step prediction values from trained GRU model
8. Calculate the error vector, mean and standard deviation of errors.
9. Product error vector
10. Estimate the threshold values based on error vector
11. do

    if SD >= threshold then
    k++
    while (timesteps)

12. if k==1 then

        Print "Warning/Point Anomaly"
    Elsif k==2
        Print "Error/Collective Anomaly"
    Else
        Print "Issue/Collective Anomaly"
    Endif

Fig. 9.6 Anomaly detection algorithm

Experiment Setup

The experiment was evaluated on Google Colab with 2 GB GPU, 16 GB RAM, and 160 GB Hard disk space, The proposed model is implemented using Python programming language with TensorFlow Keras library. Numenta Anomaly dataset is used to test the efficiency of the proposed anomaly detection model.

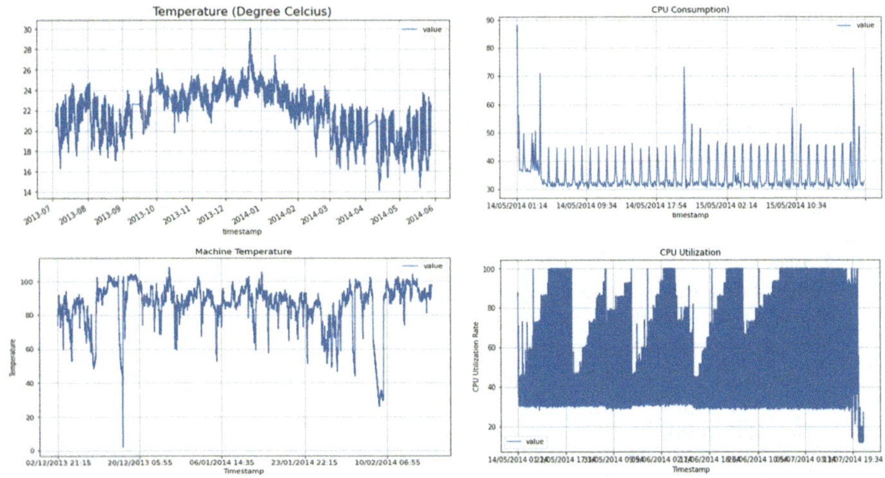

Fig. 9.7 Sample NAB dataset (i) temperature and (ii) CPU consumption values (iii) Machine temperature sensor data (iv) CPU utilization as misconfiguration

Dataset

The Numenta Anomaly Benchmark is a dataset (NAB Dataset. n.d) developed to evaluate anomaly detection algorithms in various real-time and streaming applications. The datasets in NAB (Ahmad, Lavin, Purdy & Agha 2017), namely CPU utilization as misconfiguration, ambient temperature system failure, and machine temperature system failure are used for experiment purposes. This dataset contains a timestamp and its respected values like temperature and CPU consumption etc. Data normalization is a technique used during the data preparation phase for machine learning models. The normalization process the scale the values in the range of [0,1] without obfuscating its values. Figure 9.7 shows several datasets from NAB benchmark that exhibits short and long term temporal dataset.

Hyperparameters

Eight hyperparameters are used to tune the GRU based multi-step prediction model. These hyperparameters are i) sliding-window-size (oh1), GRU-units (oh2), dropout-rate (oh3), regularizer (oh4), regularizer-rate (oh5), optimizer (oh6), epochs (oh7), activation function (oh8) and learning-rate (oh9). The most important hyperparameter are learning rate, epochs, and dropout rate because there is a direct impact on model performance.

The performance metrics

To evaluate the proposed model, the performance metrics: precision, recall, and F1-score are used. These metrics represent the accuracy and efficiency of the anomaly detection process. To calculate these metrics it is essential to understand statistical measures of the performance of binary classification: sensitivity and specificity. The various performance metric are:

(i) Precision, P, is the percentage of correctly identified values. In other words, it is a measure of correctly identified anomalies.

$$precision = \frac{TruePositive}{TruePositive + FalsePositive} \tag{9.18}$$

Recall, R, which is the number of positive class prediction out of all positive examples,

$$Recall = \frac{TruePositive}{TruePositive + FalseNegative} \tag{9.19}$$

and (iii) F1-score, F1, which is the harmonic mean of Precision and Recall, and it provides a more accurate picture of cases that were incorrectly classified than the Accuracy Metric.

$$F1Score = \frac{2 * (Precision * Recall)}{Precision + Recall} \tag{9.20}$$

The F1-score represents the balance between precision and recall and it summarizes both. The detection rule with the highest F1 as the superior anomaly detection technique. The accuracy is another important key performance parameters to evaluate the anomaly detection model. The accuracy is defined as follows

$$Accuracy = \frac{TP + TN}{TP + TN + FP + FN} \tag{9.21}$$

$$ErrorRate = \frac{FN + FP}{TP + FP + TN + FN} \tag{9.22}$$

The various loss functions are defined for machine learning models such as 'Root Mean Squared Error' (RMSE)', 'Mean Squared Logarithmic Error Loss (MSLEL)', and 'Mean Absolute Error Loss (MAE)', etc. The RMSE is the right choice to evaluate time series prediction model, and it is defined as

$$Root\ Mean\ Squared\ Error = \sqrt{\frac{1}{n}\sum\nolimits_{i=1}^{n}(y^i - y_p^i)^2} \tag{9.23}$$

The larger RMSE indicates less accuracy and least RMSE indicates highest accuracy. The various existing works are compared to the proposed model to evaluate the performance of the model. Traditional models and machine learning model for multi-time step prediction includes 'Auto-Regressive Integrated Moving Average (ARIMA)'[12], 'Hidden Markov Model' [11], 'Support Vector Machine model' [8], GRU autoencoder model [13].

9.5 Results

In this section, the results of the proposed model are discussed. This section consists of two subsections (i) results of the proposed model and (ii) comparison of the proposed model with few existing models. Prediction accuracy, anomaly detection accuracy, and RMSE are discussed. The multi-step prediction is used to find the anomalies in the proposed model. The difference between predicted values and the actual value is more than a threshold is considered as an anomaly. The GRU network models are trained with different time-step from (1 to 10) to predict the future values of time series data. Figure 9.8 shows the prediction error for multi-step predictions. The accuracy decreases as the number of time-steps are increases. To improve accuracy, it is essential to take multiple time-steps predictions. The accuracy of anomaly detection is dependent on multi-step ahead prediction accuracy. So the proposed model considers timesteps: 5. Figure 9.8 shows the multistep prediction with different epochs.

The sliding window approach is used for multi-step prediction, where various window sizes (window size: 1, 5, 10, 20) are used to predict multi-step values. The following graphs show the comparison of predicted values and test values for different window sizes. It has been observed that the prediction accuracy increases as the size of the window increases. For the multi-step prediction, n input is fed to the window to predict n + 1. The accuracy increases as more inputs fed to the window. So windows size is directly proportional to accuracy. Figure 9.9 shows the prediction accuracy for different window sizes.

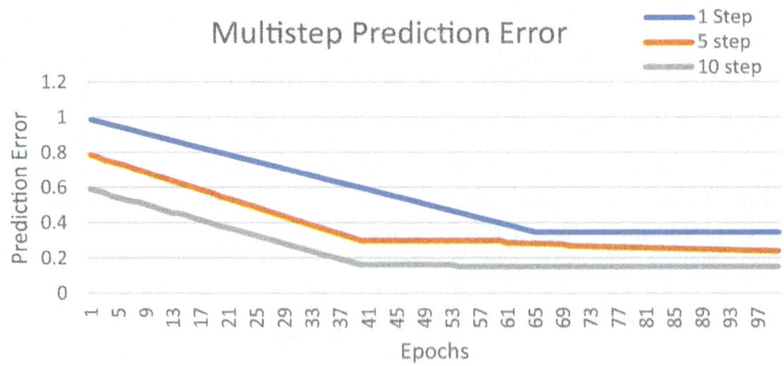

Fig. 9.8 Comparison of prediction error for various multi-step predictions

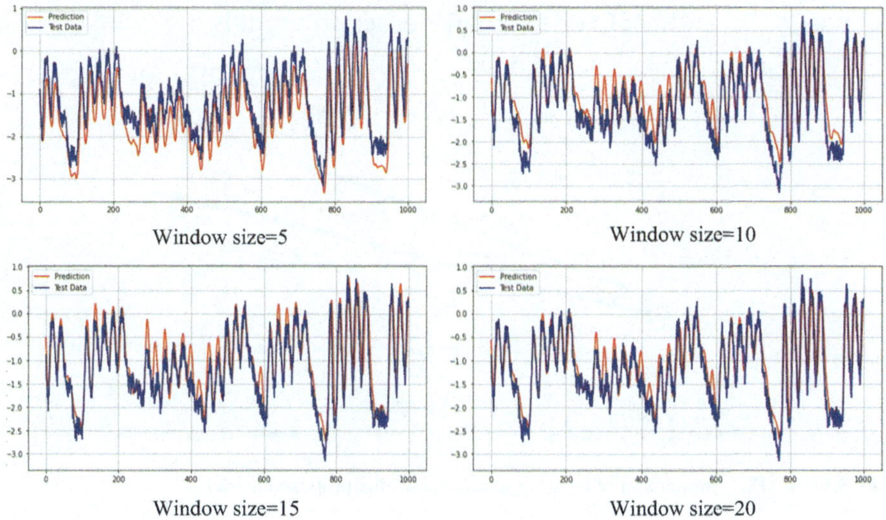

Fig. 9.9 Comparison between predicted data and test data for different window size

To evaluate the anomaly detection model, the loss function RMSE is calculated for test dataset and validation. Figure 9.10 shows the difference between train and test loss for different Epochs. The difference between test and train loss is very less, and the loss is reduced as the number of epochs in increases. From the graph, it has been observed that the accuracy of the proposed anomaly detection model is at the higher side.

Figure 9.11 shows the comparison of RMSE for various anomaly detection models. The less RMSE represents higher accuracy. The proposed model has less RMSE compared to existing anomaly detection models.

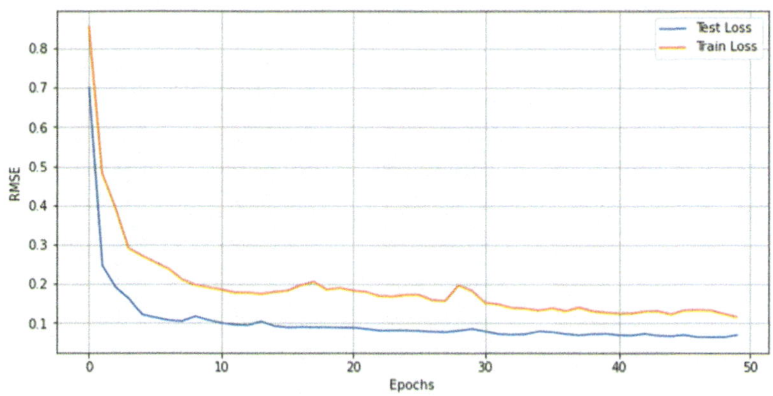

Fig. 9.10 RMSE for train and test dataset

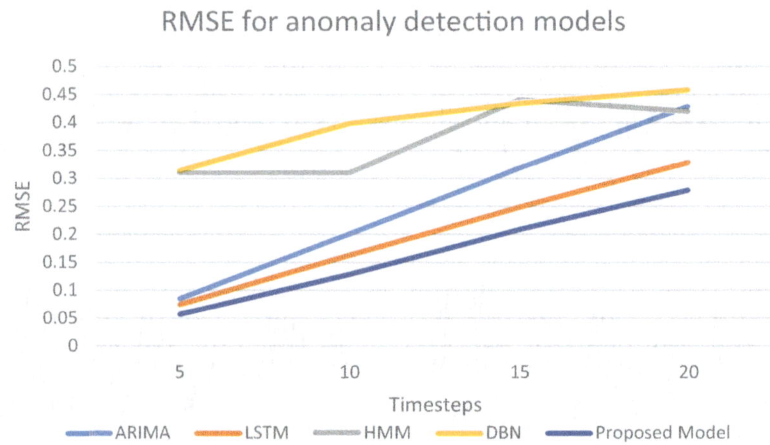

Fig. 9.11 RMSE comparison of existing model versus the proposed model

The accuracy of the proposed model has been increased compared to the existing models, but more memory and CPU cycles required for hyperparameter tuning. So proposed model can be adopted in critical applications where accuracy is the highest priority. Figure 9.12 shows the comparison of various performance parameters such as CPU consumption, memory consumption, anomaly detection time, and accuracy.

Figure 9.13 shows the recall, precision, and F1-score comparison of existing anomaly detection models with the proposed model. The existing models are results are taken from RNN, GRU [12], and auto-encoder [16]. The proposed model achieves higher precision, recall, and F1-score when compared with the other three methods.

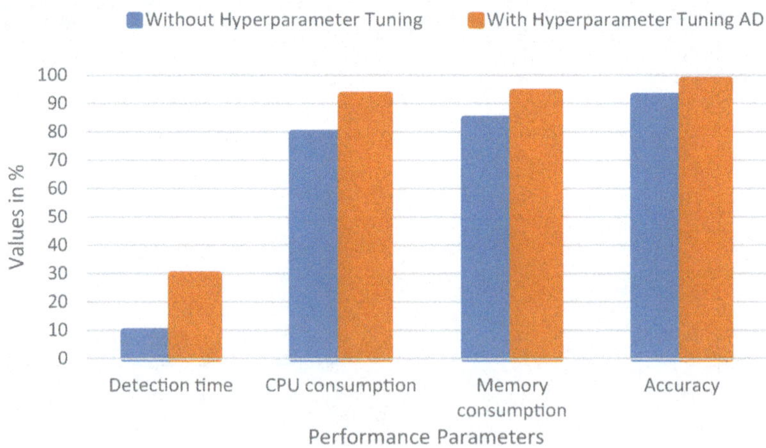

Fig. 9.12 Comparison of normal AD and hyperparameter tuner AD

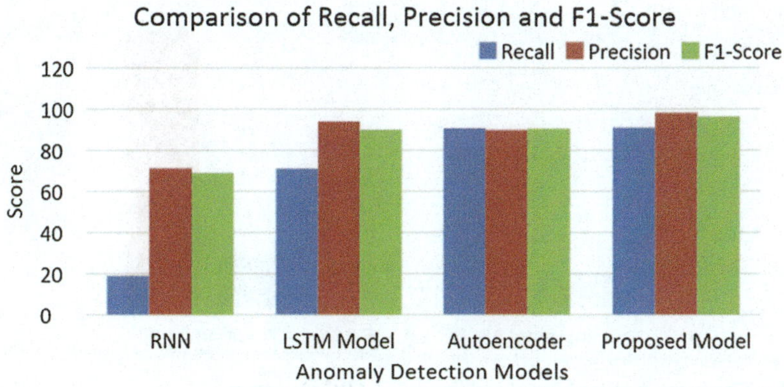

Fig. 9.13 Precision, recall, and F-score comparison of the existing and proposed system

Table 9.1 Hyperparameter values for highest accuracy

Hyperparameters	Values
Epochs	100
No. of Units	30
Dropout Rate	0.3
Regularizer	L2
Regularizer rate	0.01
Optimizer	RMSprop
Learning Rate	0.09
Activation function	ReLu

Table 9.1 shows that the hyperparameter combination with the threshold value 0.65 for which the accuracy obtained is 98.8. It has been observed from the experiment results; hyperparameter optimization gained the highest accuracy compared to other models.

Figure 9.14 shows the different between traditional and improved algorithm. During experiment evaluation, improved ABC takes less time in search and computation time.

9.6 Conclusion

Anomaly detection is one of the key aspects to improve the Quality of Service through the failure prediction. In this paper, a multi-agent based anomaly detection model is introduced to detect anomalies in the Industrial IoT. Multi agent system is used to perform various operations in the anomaly detection process. A deep learning model called the GRU model is used for multi-step prediction. The proposed model

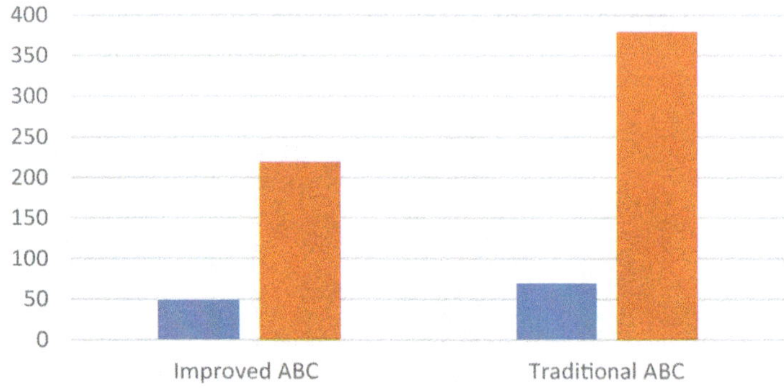

Fig. 9.14 Comparison of traditional and improved ABC algorithm

considers certain timestamp value as an anomaly if the difference between actual value and predicted value is greater than threshold value. To improve the accuracy of prediction, the GRU model adopted a bio-inspired algorithm called Artificial Bee Colony (ABC) for tuning hyperparameters. The proposed model is evaluated in the google Colab environment with the NAB temperature and CPU utilization data set. The experiment result shows that the proposed model improves accuracy compared to existing anomaly detection models. In future studies, the detection time can be improved by using new optimization techniques.

References

1. M.V. Ngo, T. Luo, H. Chaouchi, T.Q.S. Quek, Contextual-bandit anomaly detection for IoT data in distributed hierarchical edge computing, in *2020 IEEE 40th International Conference on Distributed Computing Systems (ICDCS)* (2020), pp. 1227–1230. doi: https://doi.org/10.1109/ICDCS47774.2020.00191
2. C.M. Bulla, M.N. Birje, A multi-agent-based data collection and aggregation model for fog-enabled cloud monitoring. Int. J. Cloud Applicat. Comput. **11**, 73–92 (2021)
3. M.V. Ngo, H. Chaouchi, T. Luo, T.Q.S. Quek, Adaptive anomaly detection for IoT data in hierarchical Edge Computing. arXiv:2001.03314 [cs, stat]. (2020)
4. Utomo, D., Hsiung, P.-A.: Anomaly Detection at the IoT Edge using Deep Learning. In: 2019 IEEE ICCE-TW. pp. 1–2. IEEE, YILAN, Taiwan (2019)
5. M.N. Birje, C.M. Bulla, Cloud monitoring system: basics, phases and challenges. Int. Jou. Rcnt. Tech. & Engg. **8**, 4732–4746 (2019)
6. S. Thudumu, P. Branch, J. Jin et al., A comprehensive survey of anomaly detection techniques for high dimensional big data. J Big Data **7**, 42 (2020). https://doi.org/10.1186/s40537-020-00320-x
7. J. Wu, L. Yao, B. Liu, Z. Ding, L. Zhang, Combining OC-SVMs with GRU for detecting anomalies in telemetry data with irregular intervals. IEEE Access. **8**, 106648–106659 (2020)
8. M. Birje, C. Bulla, Cloud monitoring system: a review. Int. J. Sci. Techn. Eng. Mgt. **1**(1), 49–55, (2019)
9. C. Mitchell, R. Agrawal, J. Parker, The effectiveness of edge centrality measures for anomaly detection, in *2019 IEEE International Conference on Big Data (Big Data)* (2019), pp. 5022–5027. doi: https://doi.org/10.1109/BigData47090.2019.9006468

10. S. Maya, K. Ueno, T. Nishikawa, dGRU: a new approach for anomaly detection using deep learning with delayed prediction. Int. J. Dat. Sci. Ana. **8**, 137–164 (2019)
11. Y. Tan, C. Hu, K. Zhang, K. Zheng, E.A. Davis, J.S. Park, GRU-based anomaly detection for non-linear dynamical system. IEEE Access. **8**, 103301–103308 (2020). https://doi.org/10.1109/ACCESS.2020.2999065
12. Y. Gao, An improved hybrid group intelligent algorithm based on artificial bee colony and particle swarm optimization, in *2018 International Conference on Virtual Reality and Intelligent Systems (ICVRIS)* (2018), pp. 160–163. doi: https://doi.org/10.1109/ICVRIS.2018.00046
13. M. Said Elsayed, N.-A. Le-Khac, S. Dev, A.D. Jurcut, Network anomaly detection using GRU based autoencoder, in *Proceedings of the 16th ACM Symposium on QoS and Security for Wireless and Mobile Networks*, pp. 37–45 (2020)
14. B. Laughlin et al., A service architecture using machine learning to contextualize anomaly detection. J. Database Manag. **31**(1), 64–84 (2020). https://doi.org/10.4018/JDM.2020010104
15. A. Alghuried, *A Model for Anomalies Detection in Internet of Things (IoT) Using Inverse Weight Clustering and Decision Tree* (Technological University, Dublin, 2017)

Chapter 10
Nurturing the Rudiments and Use Cases of Ongoing Natural Language Generation for a Future Profitable Business More Profitable

R. Sivarethinamohan and S. Sujatha

Abstract Decoding the world of artificial intelligence and its usage in the current intelligence landscape enhance bottom-up growth in building resilient global business. The areas of artificial intelligence (AI) concerned with human-to-machine and machine-to-human interaction. The Next Wave in AI-driven speech is Natural Language Generation (NLG). Natural Language Generation (NLG) is the use of artificial intelligence (AI) programming to produce written or spoken narrative from a dataset. NLG is related to computational linguistics, natural language processing (NLP) and natural language understanding (NLU). NLG research often focuses on building computer programs that provide data points with context. Sophisticated NLG software has the ability to mine large quantities of numerical data, identify patterns and share that information in a way that is easy for humans to understand. The speed of NLG software is especially useful for producing news and other time-sensitive stories on the internet. At its best, NLG output can be published verbatim as web content. The goal of Natural language generation (NLG) is to use AI to produce written or spoken narrative from a dataset. Therefore, this study aims to study how NLG enables machines and humans to communicate seamlessly, simulating human to human conversations and using NLG how organizations are building new customer experiences, monetizing information assets, introducing new offerings and streamlining operational costs. Therefore, the coverage of this chapter will answer to the industrialists and new start-ups. What can NLG do for business? and what are the future applications of NLG?

Keywords Lexicalisation · Sentence aggregation · Orthographic realization · Micro planning · Dialogue system

---

R. Sivarethinamohan (✉)
Department of Professional Studies, CHRIST (Deemed to be University), Bengaluru, India

S. Sujatha
K Ramakrishnan College of Technology, Trichy, Tamilnadu, India

S. L. Fernandes and T. K. Sharma (eds.), *Artificial Intelligence in Industrial Applications*, Learning and Analytics in Intelligent Systems 25,
https://doi.org/10.1007/978-3-030-85383-9_10

10.1 Introduction

Language is the paramount mechanism for moving information between brains. In fact, the principle of linguistic relativity (AKA the Sapir-Whorf hypothesis) strongly advocates that humans depend on language in order to think. If they are to ever develop models that encapsulate the full complexity of human thought, they will need them to understand language [3]. Advancements in deep learning algorithms and the availability of sufficient computational power have brought NLG for the automatic generation of human-readable text by a computer [4]. For example, NLG is used by the Associated Press to automatically generate business earnings reports. Every week, The Washington Post uses Heliograf, their in-house automated storytelling technology, to cover all high school football games in the Washington, D.C. area. Small businesses deal with a large number of consumer inquiries and are unable to respond to all of them. However, with NLG, a chatbot may respond to these questions. If a consumer has to wait 15 to 20 min to interact with an agent, they may prefer to chat with a bot. Anyone receiving information from a well-trained NLG model can gain a competitive advantage. Therefore, this chapter can help CIOs, analysts, compliance officers, portfolio managers, or even the general public.

10.2 Rudiments of Natural Language Generation

Natural Language Generation is a subclass of Natural Language Processing and is the art of generating text from structured data using AI. In other words, any text embedded as a part of a document, report, explanation, or any other help message within a speech stream can be used as the input, while the output can be any non-linguistic representation of information. Any communicative database can be used as a knowledge source for the NLG. Figure 10.1 depicts how does NLG work?

The NLG software converts structured data into human-readable language by following a few simple steps. Figure 10.2 shows software processss of NLG.

Fig. 10.1 Understanding the workings of NLG

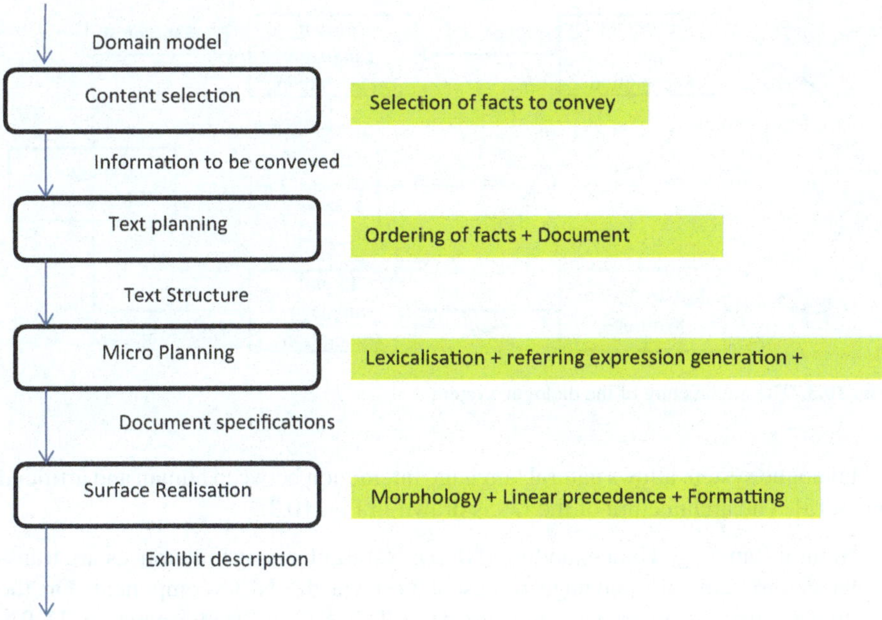

Fig. 10.2 Software process of NLG

Content determination—This allows a conversational AI to select how to react; the process is based on extensive training data that clearly describes potential intent and entity sets.

Discourse planning—This is necessary in order to organise the texts using conceptual grouping and rhetorical links.

Sentence aggregation—By mixing messages in a sentence plan, this is done to build vast, complicated sentences.

Lexicalisation—This is done to detect domain-centric jargons, depending on the developer's goal.

Referring expression generation—This section includes reference usage, including nouns, pronouns, definite descriptions, and so on.

Syntactic and morphological realization—At this point, the system is provided with basic grammar ideas including morphology and syntax.

Orthographic realization Finally, proper punctuation and case, as well as typographic criteria such as font size, are important for language production.

10.2.1 Understanding the Working of NLU and NLG

Based on the dialogue system, let us describe the distinction between natural language understanding (NLU) and natural language generation (NLG).

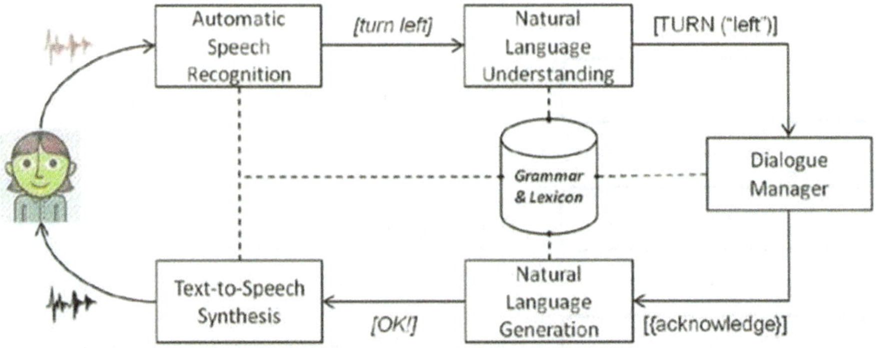

Fig. 10.3 The architecture of the dialogue system

Dialogue system allows natural language interaction between human and artificial agent [5]. The architecture of the DS is drawn in Fig. 10.3.

- Natural Language Understanding (NLU): Natural language utterances are translated into semantic language representation via the NLU component. On the image "Turn left" utterance is translated into TURN ("left") representation. TURN ("left") can be expressed in a variety of ways, such as "I must turn left," "Please take left," and so on. Natural language variability (where the same meaning can be represented in many ways) is dealt with by the NLU component, which establishes many-to-one correspondence.
- Natural Language Generation (NLG): NLG component does the opposite translation from the semantic representation into natural language utterances. For example, [acknowledge] might be translated as "OK!" instead of "OK!" It could also be "roger," "understand," "no issue," and so forth.

10.2.2 Language Modeling Toolkit for NLG

Language modeling toolkit are most popular resources used in natural language generation in machine Transaction. Some of the popular language modeling toolkit are.

(i) CMU language modeling tool:- The Carnegie Mellon Statistical Language Modeling (CMU SLM) Toolkit is a collection of unix software tools aimed to help researchers with language modelling.
(ii) SRILM: It is a toolkit for creating and using statistical language models (LMs) in speech recognition, statistical tagging and segmentation, and machine translation. SRILM runs on UNIX and Windows platforms.
(iii) Microsoft language modeling tool:- Microsoft Research uses MSRLM as its internal language modeling tool chain. It was built to handle massive amounts

of data. Infact, In less than eight hours, a language model was developed on cutting-edge hardware using 40 billion words of web data.

(iv) Active learning and domain adaption methodologies can be used in order to employ a target language corpus for NLG [6].

There is currently no end-to-end NLG library or module available. Users can make use of those libraries to assist them in writing their content selection or surface realisation modules. There is no module for the stage of discourse/document planning or microplanning. For surface realisation, users who utilise Java can use simple NLG or open CCG. They must write their own module for the planning step. They can use those NLP functions for lexicalization. A library in Python contains a large number of lexical transformation functions. Pattern is also useful for their other NLP requirements [13]..

10.2.3 Preferred Programming Language for NLG

Python is the best. SciPy and NumPy make it more powerful. Then there are the python-friendly Theano, Dis-sect, and Gensim.

Java comes in second. People are familiar with it because it is extensively utilised. Many universities make open source Java tools and libraries available [7].

10.2.4 NLG Libraries to Detect Topics from a Given Text

The following tutorials/tools assist users in identifying from a given text:

(i) Topic Modelling in Python with NLTK and Gensim—Towards Data Science
(ii) Complete Guide to Topic Modeling—NLP-FOR-HACKERS
(iii) Stanford Topic Modeling Toolbox
(iv) Topic Modeling.

10.2.5 List of Requirements for Effective Natural Language Generation

- The structure in which the information is conveyedon.
- The flow of sentences in a coherent, easily comprehensible way
- The choice of words to convey the information without misunderstood or misinterpreted
- The right referring expressions(anaphoras, exophoras and other kinds of pronouns)

- The syntax and language-specific conventions that ensure to the syntax and conventions to that language.

10.2.6 Different Variations of NLG

(i) Basic NLG: Basic NLG uses Excel-like routines to automatically translate dats into text. A mail merge that restates numbers into language is an example of this.

(ii) Templated NLG: The user creates templates, decides how to connect concepts, and interprets the results. To automate content specifically in sentence building, it relies on business rules, basic calculations (for example, sum), and templates with boilerplate language. With no grasp of what the user is seeking to say or their specific domain, templated systems are limited in their capacity to pull from numerous data sources, do advanced analytics, achieve reusability from one project to the next, and explain how it came to the narrative it developed.

(iii) Advanced NLG: Advanced NLG communicates in the same way as humans do, for example, by incorporating intelligence and intent from the start. It analyses data to determine what is significant and interesting to a certain audience, then converts those insights into Intelligent Narratives, such as smart messages packed with audience-relevant information written in conversational language.. Backed by a knowledge base, Advanced NLG systems understand the domain and can write contextually about a user's business at a scale that is not possible by humans.

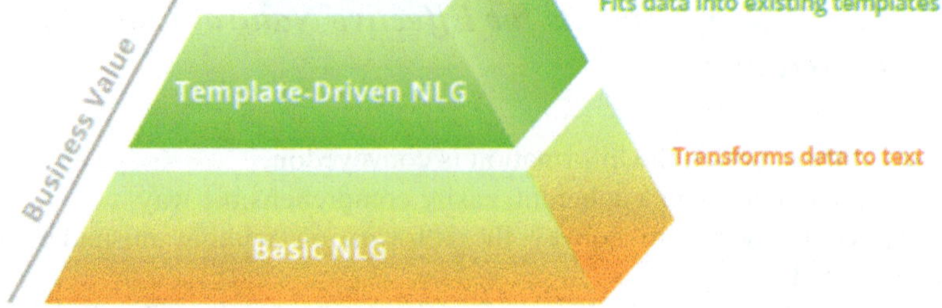

10.3 Use Cases of Natural Language Generation

We now discuss how targeted NLG applications and solutions sustain profitable business.

Business chatbots:

Many chatbot technologies rely on NLG to partially or completely create natural language responses to user questions. NLG is used by advanced chatbots to interpret what users input and then produce responses based on what they say. This could involve customizing language to the given request or accent, or providing targeted messaging based on the topic. It could also entail gathering content from various sections of the website and packaging it into a response to specific questions posed by users. It could also entail gathering content from various sections of the website and putting it into a response to particular questions you've posed.It's useful in places like cybersecurity articles, white papers, and research. It's also used to perform sentiment analysis on web material in order to improve customer service and present better solutions [1].

Entertainment bots:

These days, entertainment bots are all the rage. They can strike up a casual discussion with the user about whatever they like. Even task-specific bots (such as those for Pizza Hut or CNN) should be able to speak in a human-like manner with users or consumers. The idea is for users to feel as if they are conversing with a human when speaking with a bot. That is the goal of NLG systems.

Emails:

NLG is widely used in Gmail, where it generates automated responses. It's also used to generate reports from large amounts of data, such as narrative descriptions of company statistics and charts.

Voice Assistants and Digital assistants:

NLG is used by major voice assistants such as Apple's Siri and Amazon's Alexa to answer questions and requests from users. NLG is used by any software application or gadget that responds when a user talks to it. A digital assistant, on the other hand, seems to be a more advanced sort of chatbot that can handle more complicated interactions in a conversational manner.

Automated Lead Nurturing:

NLG is used by several automated lead nurturing platforms. These are AI-powered apps that use NLG to automatically respond to leads via email and chat. This ensures that all leads entering a pipeline receive a response 24 h a day, seven days a week, whether or not human reps are accessible.

Protection and Health Insurance:

NLG can analyse data and generate reports on individual customers. It gives a comprehensive view of the risks and the premiums that should be charged. It speeds up underwriting with due care on quality and customer care. Therefore, It can reduce underwriting time.

Stock market reporting:

NLG can analyse data and inform as to the best-performing and worse-performing stocks, their performances, and the respective indices on which those stocks are listed.

Financial reporting:

Figures such as EBIT, EBITDA, Revenue, Operating Expenses, and Net Income may all be identified, converted to text, and placed in the right historical perspective. In the realm of financial reporting, natural language generation systems are also adaptive, which means they can produce a variety of styles and types of reports(Financial Reports, Regulatory Filings, Executive Summaries, Suspicious Activity Reports, Strategic Advice,Breaking News And Chart Annotations), each reflecting the unique needs of specific target groups.

Beratungsdokumentation (consulting documentation):

The beratungsdokumentation is a necessary form for advisers that explains why the advice they are offering to consumers is appropriate for them. Further it is one of the regulations under Mifid II. NLG might be used to automatically create this type of document.

E-commerce:

NLG technology is used by e-commerce suppliers to automate product listings and descriptions. The benefit of NLG in this area is that it allows user to change entire texts in the blink of an eye by just switching or adding datasets.

Marketing campaign:

When creating a marketing campaign, users must not only target the correct people at the proper time and place, but there is another crucial component to consider (the right message). What are the possibilities that the user will choose the right words and linguistic qualities to engage the audience?

Image Captions:

Breakthroughs in computer vision, natural language processing, and deep learning algorithms have made it possible to automatically produce captions from photos. Image caption is a key aspect of scene understanding, which combines computer vision and natural language processing knowledge to automatically generate natural language descriptions based on the content observed in an image [8].

Analytics Dashboards:

Dashboards for analytics are one of the first and most visible applications of natural language generation. To make rapid and successful judgments, business executives frequently demand information in a readily understandable format. They frequently lack the time to go through pages of analytics data or examine a slew of charts and graphs in order to extract useful information. As a result, NLG-powered tools can be utilised to evaluate data from analytics in compact yet comprehensive reports generated from the data acquired. Leaders may always have the most up-to-date information for making critical choices using NLG-powered analytics dashboards [9].

Automated Journalism:

A good example of NLG is automated journalism. Where a computer searches the web for real-time news and scapes the data from different sources and writes a text summary that can be published very quickly to the web.

Other potential applications

(a) Analysis for business intelligence dashboard
(b) IoT device status and maintenance reporting
(c) Reporting on business data/data analysis
(d) Individual client financial portfolio summaries and updates
(e) Personalized computer communication
(f) Understanding the human language production facility
(g) Planning extended monologic discourses
(h) Psycholinguistic modelling
(i) Composing target texts in machine translation systems[10].

10.4 The Best Natural Language Generation Platforms

The best Natural Language Generation Platforms are:

a. Arria NLG:

Arria NLG is a part of artificial intelligence that converts structured data into natural language. Arria software replicates the human process of expertly understanding and delivering data insights through data analysis, knowledge automation, language generation, and targeted information delivery. Arria extracts and summarises data into understandable narratives using NLG.

b. Amazon Polly:

Amazon Polly is an API-driven service that synthesises speech that sounds like a human voice using powerful deep learning capabilities. Polly turns text into natural-sounding voice, allowing users to construct talking apps and create entirely new categories of speech-enabled products. Polly's TTS (Text-to-Voice) is a service that synthesises natural-sounding human speech using powerful deep learning technologies [2].

c. AX Semantics:

AX Semantics is a natural language generation (NLG) software firm that uses AI.Its SaaS-based platform allows ACG to serve consumers of all sizes and is extensively utilised in the e-commerce, business, banking, and media publishing industries. AX Semantics, which is available in 110 languages, contains capabilities such as structured data-based NLG output, multi- and cross-language text production, an intuitive inline editor, and more.

d. Automated Insights:

Wordsmith, an NLG technology that turns data into intelligent storey, was created by Automated Insights. Complete narrative customisation, real-time content updates, and a strong API for flexible publishing are all possible with the platform.

e. Quill:

Quill is an intelligent automation technology that uses NLG to assist businesses transform their reporting procedures. It converts data into tales, which are then embedded straight into dashboards. Quill also allows users to automatically build data-driven stories at scale.

f. Textengine.io

Textengine.io is a self-service platform for NLG. The platform is based on artificial intelligence and translates structured data into text automatically. Intelligent language analysis, an intuitive interface, automatic, machine translation of text structures, and other capabilities are included.

g. vPhrase

vPhrase Analytics is an AI firm that uses machine learning and natural language generation technology to give business intelligence and automated reporting.

h. Yseop

Yseop offers a variety of Natural Language Generation solutions to support report writing, sales management, and other business activities by combining human experience with powerful artificial intelligence technologies.

10.5 Future of Natural Language Generation

Conversational interfaces, virtual assistants, and software applications are all areas where NLG has a bright future (like weather forecasting, news writing, canned text generation etc.). Alexa, Cortana, and other intelligent personal assistants are entering into a new era of convenience and efficiency for customers. With conversational interfaces that facilitate engagement across employees and with customers, the enterprise is catching up, raising the bar on how these technologies communicate [12].

10.6 The Major Challenges on Natural Language Generation (NLG)

NLP is a strong technology with numerous advantages, however there are still a number of limitations and issues with Natural Language Processing:

a. Contextual words and phrases and homonyms
b. Synonyms
c. Irony and sarcasm
d. Ambiguity
e. Errors in text or speech
f. Colloquialisms and slang
g. Domain-specific language
h. Low-resource languages
i. Lack of research and development [14].

10.7 Conclusion

Present-day NLG applications cover a range of use cases in areas such as journalism, business and financial reporting and the writing of product descriptions. For the marketer, the technology makes the writing of short-form marketing copy more effective. In the case of email marketing, for instance, NLG can write subject lines that are tailored to a brand's audience, and optimized for maximum engagement. NLG systems can adapt the tone, style and structure of the output language to a brand's audience, context, and narrative purpose [11]. It is a complicated task that takes into account multiple aspects of language such as grammar, syntax, sentiment, word usage, and perception that can ensure the peaceful co- existence of all beings on the planet.

References

1. E. Adamopoulou, M. Lefteris, Chatbots: history, technology, and applications. Mach. Learn. Appl. **2**, 100006 (2020). https://doi.org/10.1016/j.mlwa.2020.100006
2. A. Choudhury, Analytics India magazine: Top Natural Language Generation Platforms in 2021 (2021). https://analyticsindiamag.com/top-natural-language-generation-platforms-in-2021/

3. J. Bateman, M. Zock, Natural language generation. Oxford Handbooks Online (2012). https://doi.org/10.1093/oxfordhb/9780199276349.013.00
4. A.J. Cawsey, B.L. Webber, R.B. Jones, Natural language generation in health care. J. Am. Med. Inform. Assoc. **4**(6), 473–482 (1997). https://doi.org/10.1136/jamia.1997.0040473
5. R. Dale, Natural language generation: the commercial state of the art in 2020. Nat. Lang. Eng. **26**(4), 481–487 (2020). https://doi.org/10.1017/s135132492000025x
6. O. Dušek, J. Novikova, V. Rieser, Evaluating the state-of-the-art of end-to-end natural language generation: the E2E NLG challenge. Comput. Speech Lang. **59**, 123–156 (2020). https://doi.org/10.1016/j.csl.2019.06.009
7. E. Reiter, Natural language generation, in A. Clark, C. Fox, S. Lappin (eds.), Handbook of Computational Linguistics and Natural Language Processing, John Wiley & Sons 574–598 (2010)
8. E. Reiter, R. Dale, Building natural language generation systems (Studies in Natural Language Processing) Cambridge University Press, Cambridge, UK, 1st edition (9 March 2006)
9. A. Mykowiecka, Natural-language generation—an overview. Int. J. Man Mach. Stud. **34**(4), 497–511 (1991). https://doi.org/10.1016/0020-7373(91)90032-3
10. E.Reiter, R. Dale, Building applied natural-language generation systems. J. Nat.-Lang. Eng. **3**, 57–87 (1997)
11. R. Sivarethinamohan, S. Sujatha, Unraveling the potential of artificial intelligence-driven blockchain technology in environment management, in G. Manik, S. Kalia, S.K. Sahoo, T.K. Sharma, O.P. Verma (eds) Advances in mechanical engineering. Lecture notes in mechanical engineering. Springer, Singapore (2021). https://doi.org/10.1007/978-981-16-0942-8_66
12. R. Sivarethinamohan, D. Yuvaraj, S. Shanmuga Priya, S. Sujatha, Captivating profitable applications of artificial intelligence in agriculture management, in P. Vasant, I. Zelinka, GW Weber (eds) Intelligent computing and optimization. ICO 2020. Advances in intelligent systems and computing, vol. 1324. Springer, Cham (2021). https://doi.org/10.1007/978-3-030-68154-8_73
13. M.O. Topal, A. Bas, I.V. Heerden, Exploring transformers in natural language generation: GPT, BERT, and XLNet (2021). ArXiv, abs/2102.08036
14. K. Van Deemter, E. Reiter, H. Horacek, Formal issues in natural language generation. Res Lang Comput **4**, 1–7 (2006). https://doi.org/10.1007/s11168-006-9008-9

Chapter 11
A Study on Deep Learning Models for Medical Image Segmentation

Pashupati Bhatt, Ashok Kumar Sahoo, Saumitra Chattopadhyay, and Chandradeep Bhatt

Abstract In this era, Deep Learning becomes very popular in the medical science domain, especially for the segmentation of medical images. Prediction of the symptoms of dangerous diseases now becomes possible due to Machine Learning and its subset algorithms. Deep learning, a variant of machine learning in Artificial Intelligence, has features capable of learning from unlabeled data. Currently, image segmentation gets a boost and attention from researchers across the world. Image segmentation means an image gets break down into small regions having different characteristics and provides the meaningful information/objects of the image. Various image segmentation techniques are available, which divides the image into multiple parts. This division depends on image features like pixel color, intensity, texture, etc. There is a categorization of techniques base on the segmentation method used. This review chapter highlights the most popular algorithms based on DL architecture with respect to segmentation of medical images. The review comes to the point that DL is still a demanding technology especially for medical areas where bio-medical images need to be analyzed properly and more accurately.

Keywords Artificial intelligence · Medical imaging · Natural language processing · Deep learning · Medical image segmentation

11.1 Introduction

If we go a few years back, it has hardly dreamed up that deep learning applications make it possible like auto-driving cars and virtual assistants like Alexa, Siri, and Google Assistant. However, today, these tools and applications are a part of our daily life.

P. Bhatt · A. K. Sahoo (✉) · C. Bhatt
Department of Computer Science & Engineering, Graphic Era Hill University, Dehradun, India

S. Chattopadhyay
Dehradun, India

S. L. Fernandes and T. K. Sharma (eds.), *Artificial Intelligence in Industrial Applications*, Learning and Analytics in Intelligent Systems 25,
https://doi.org/10.1007/978-3-030-85383-9_11

Today we see that there is a huge involvement of Deep Learning (DL) algorithm [1] in research field such as natural language processing (NLP), computer vision, healthcare, entertainment and speech analysis, etc. This method best fits those areas, where the intelligence level is just like an expert human. The role of DL in computer vision as a pattern identification or image segmentation tool is now becoming a significant point within the field of medical image analysis.

Medical imaging sector [2] day by day is moving forward by using the advance computer technology like advance hardware design, safe and secure procedures, and latest computational resources and data storage capabilities etc. Images generated for diagnosis purpose through various imaging technique are high-resolution images, needs segmentation, classification, and abnormality detection as fast as possible with high accuracy.

This requires a great change within the development of medical image processing algorithms. So that focusing on the automation process for image analysis and fetching useful information is very crucial.

An algorithm implemented in digital computer used for processing the image known as digital image processing [3]. Image segmentation is a challenging and main process of image processing. Image segmentation means partition of an image into several meaningful parts having similarities.

DL whose main characteristics is automatic feature generation without human involvement with reliable performance has compelled researchers to utilize this strength in medical science to forecast or detect dangerous diseases.

In this COVID-19 pandemic time [4], there is an urgent requirement to get the covid-19 test report as fast as possible. Many researchers are working on this topic, using DL techniques for the analysis of medical images and disclosed significant results.

A lot of research works done using DL techniques are a readily available as an open-source model, on various online platforms. This make possible to enhance the earlier work rapidly and quickly applying new techniques to resolve medical image segmentation problem. The challenges are still there because the person who is DL experts generally has a piece of limited clinical knowledge and vice versa.

The rest part of this chapter contains a brief overview of medical imaging, image segmentation, and some popular deep learning models. Finally a review of some chapters where DL algorithms are the prime factor for medical image segmentation (MIS).

11.2 Medical Imaging

Medical imaging [2] means a techniques and processes by which we can make visual knowledge of the various inner parts of the human body in the form of images. So that these images can be used for diagnostic and treatment purposes. Medical imaging is paly important role in modern healthcare systems. Several types of medical images are there and they dependent on the method of imaging technique used to create it. Some popular techniques are below.

- X-Ray Imaging
  X-ray (electromagnetic radiation) imagings develop 2D pictures of the inside of our body and show the parts of our body with various shades of black and white. Its main application is to detect fractures in bone.
- Computer Tomography (CT) Scan
  CT scan used in radiology to get detailed images of the body with the help of computer and rotating X-ray machines for diagnostic purposes. CT scan images [5] have more details and are better than X-rays images.
- Magnetic Resonance Imaging (MRI)
  With the concept of the electromagnetic field and a specific type of electromagnetic radiation (radio waves) in radiology, MRI is a popular Imaging technique for creating high-quality images from insides of the human body mainly for organs and tissues.
- Ultrasound Images
  In the absence of any radiations and using sound waves there is an imaging technique called ultrasound or sonogram, which is perfectly fit to get an image of tissues, organs, blood flow, and other structures inside the human body. Mainly used to monitor the fetus during pregnancy.
- Positron Emission Tomography (PET)
  PET scan is an effective way of imaging for the detection of early signs of cancer, brain disorder, and heart disease.

11.3 Segmentation of Image

Several image segmentation techniques [6, 7] are in exist, which partition the image into multiple meaningful parts so that it become easier to analyze. The criteria of partition depends on image features like pixel intensity, color, texture, etc. The popular techniques used for image segmentation are shown in Fig. 11.1. The image segmentation technique selection depends on the type of images and application areas.

11.4 Deep Learning Overview

Deep Learning [8] comes under ML and ML comes under AI. A mathematical relationship among DL, ML, and AI can be represented like—DL $\subset$ ML $\subset$ AI. AI is a very popular technology using today to make it possible that a machine can show intelligence like a human to some extent. The overview of DL, ML and AI is shown in Fig. 11.2.

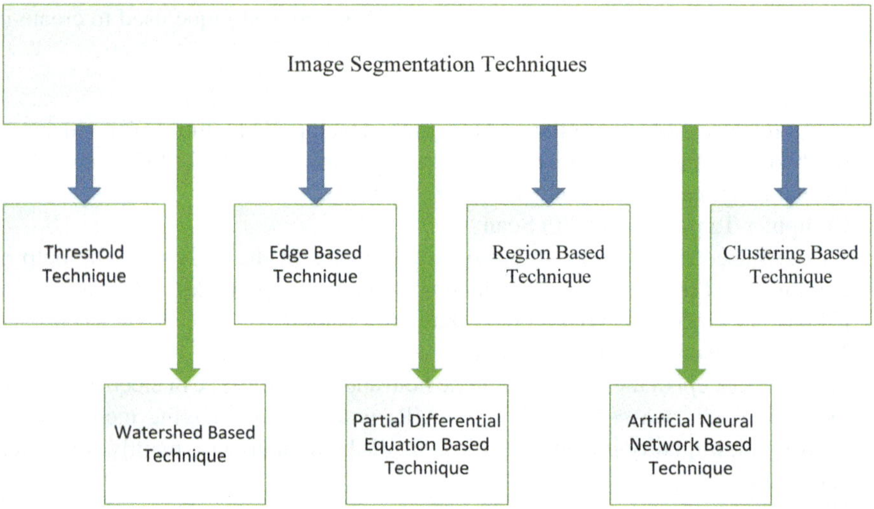

Fig. 11.1 Image segmentation techniques

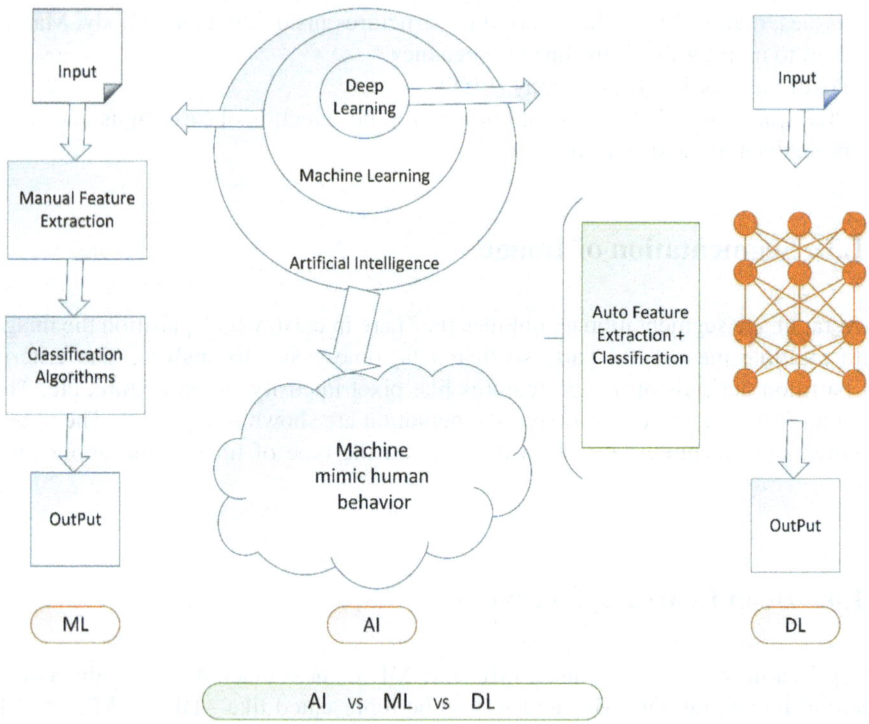

Fig. 11.2 Overview of deep learning

ML is an approach to achieve AI through a set of programs (algorithms), which get trained through featured data provided manually. DL is special sub part of ML totally based on the concept of biological neurons of the human brain. With respect to DL, this multi-layer concept of algorithms called ANN or simply neural network. Automatic feature extraction is the main feature of DL which makes it better than ML. Deep learning represents a conceptual shift in thinking, from "How do you extract the best features?" to "How do you guide the model toward discovering the best features?".

11.4.1 Deep Learning Architecture

All Deep Learning networks [9] derived from the basic network simply called ANN (or neural network). In this architecture, there is a mutually joined group of nodes, each connection is similar to the synapses in a biological brain and can transfer signal to other nodes (neuron). In Fig. 11.3, artificial neuron represented by circular node and connection between two node represented by an arrow.

Some of the widely used architectures in medical image segmentation are depicted below.

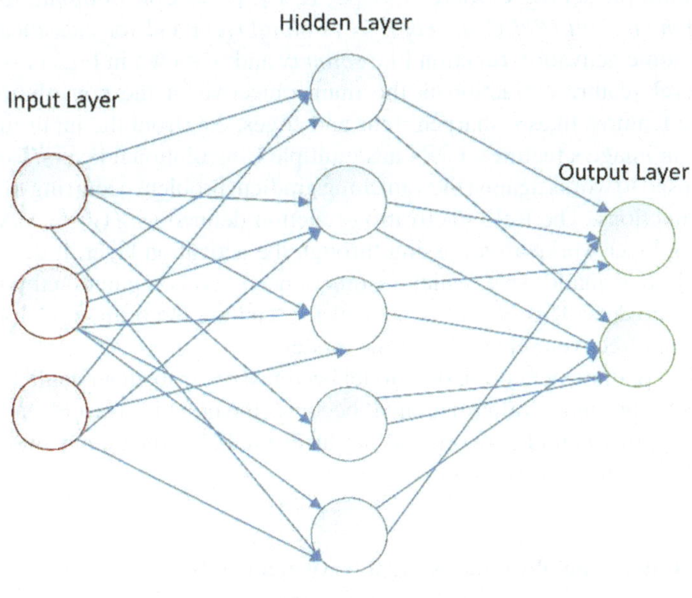

Fig. 11.3 Artificial neural network architecture

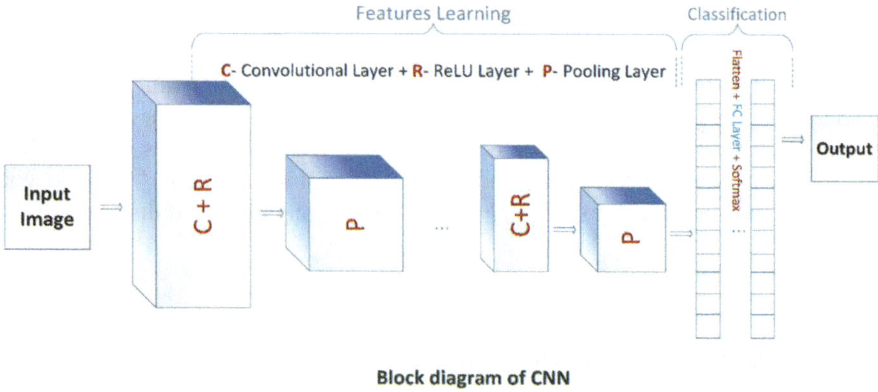

Fig. 11.4 The convolutional neural network

11.4.1.1 Convolutional Neural Network (CNN)

CNN (or known as ConvNet) [10] is the most widely used algorithm in pattern recognition and image processing. CNN is a combination of multiple layers for performing two main task i.e. (a) Feature learning (b) Classification. For first task there is a multiple set of **A** where A = {C, R, P}. [*C* = *Convolutional layer*, *R* = *Rectified Linear Unit (ReLU) layer*, *P* = *Poolinglayer*] and for classification aFC layer with some activation function like *softmax* and is shown in Fig. 11.4.

High-level feature extraction is the main objective of the convolution operation. These features likes—sharpen, blur and edges, etc. from the input image. For extracting an image's features, CNN has multiple Convolutional layers. ReLU layer in CNN is used to works against the vanishing gradient problem occurring in sigmoids and tanh functions. The ReLU activation function defined as: $f(g) = \text{MAX}\ (0, g)$. The pooling layer works after passing through the activation layer. It decreases the image stack to a smaller size, which results to need less computational potential to manipulate the data. This is very beneficial for fetching the primary characteristics necessary for effective preparation of the model.

After final pooling operation, our model becomes ready to understand the features of the object. This output (after flattened) becomes the input fora last FC layer, which is mostly a regular neural network and produces final classification using activation functions like s*oftmax* or *sigmoid.*

11.4.1.2 Fully Convolutional Neural Network (FCN)

Deeper layers of CNN have good "semantic information" but poor "spatial information". CNN is designed to be spatially invariant, that is they are not sensitive to the position of an object in the picture. To overcome this problem last fully connected (FC) layer of the CNN are modified. A new transposed convolutional layer acts in

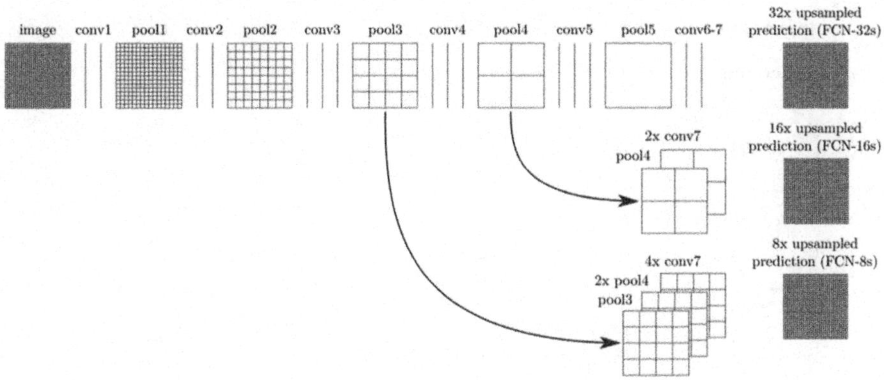

Fig. 11.5 Fully convolutional network

place of the FC layer. It uses up sampling to get a larger size of an image by recovering the initial spatial dimensions. Therefore, FCN maintains both semantic as well as spatial segmentation information [6] and is shown in Fig. 11.5.

11.4.1.3 Auto-Encoder

Auto-encoders are feed-forward unsupervised neural network where input and output are homogeneous. Mainly used for—feature extraction, space dimensionality reduction and generative modeling [11].

An auto-encoder which is shown in Fig. 11.6 has mainly two parts:

(a) Encoder—compress the input data so that similar data come together (latent space representation).
(b) Decoder—reconstruct the output from this code.

The occurrence of incoherence in latent space makes its uses limited. Various modified version are available as per requirement. Ex-Convolutional Autoencoder (CAE), The Sparse Auto-encoder, Contractive Auto-encoder, Variational Autoencoder (VAE), and Adversarial Autoencoder (AAE).

11.4.1.4 Recurrent Neural Network

Another type of neural network that is specially used for time series data is a Recurrent Neural Network (RNN) [12], which is shown in Fig. 11.7. It overcomes the drawback, which occurs in feed-forward network by handling sequential data, taking current input and pre received inputs with help of internal memory. The main application of RNN are—Image Captioning, Time Series Prediction-like predicting the prices of stocks in a particular week, Natural Language Processing, Machine Translation etc. They are mainly useful where output depends on previous results.

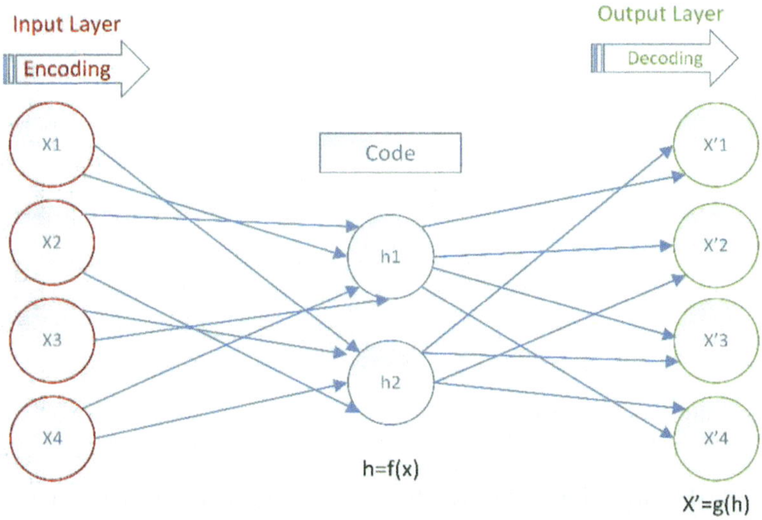

Fig. 11.6 An auto-encoder

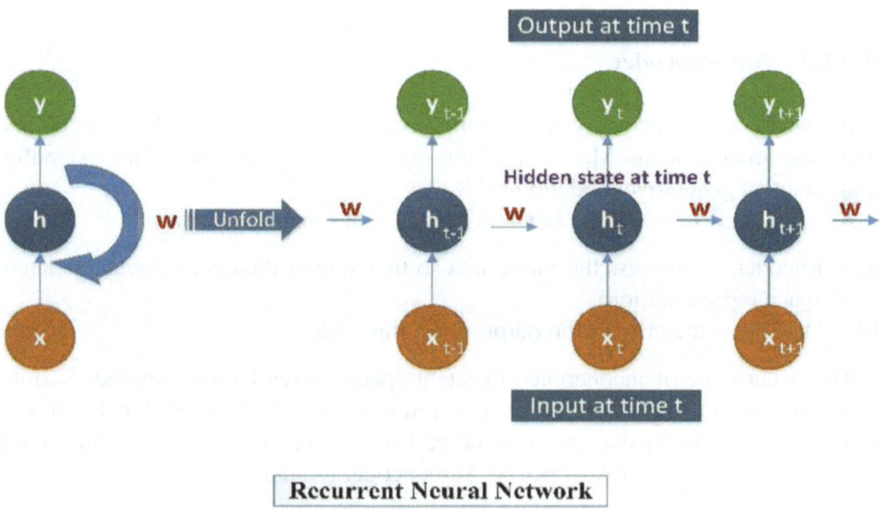

Fig. 11.7 Working of recurrent neural networks

RNNs have some learning issues due to gradient problem. When the gradient tends to small then difficulty arises to update the parameters resulting learning of long data sequences becomes difficult.

11.4.1.5 Boltzmann Machines

In category of unsupervised DL model, Boltzmann Machines (BM) [13] is an unsupervised architecture of symmetrically connected node that makes stochastic output. Boltzmann Machine is a generative deep learning model. Boltzmann Machine consists of visible and not visible nodes. It is an energy-based model and uses the Boltzmann Distribution to describe the different states of the system. In Boltzmann, machine weights of synapses refer to the energy of the system and after weights are set and the system is trained, then the system automatically tries to adjust in the lower energy state by modifying the weights.

Different types of Boltzmann Machines, as depicted in Fig. 11.8, are available. For example, RBM-Restricted Boltzmann Machines-In a BM, each node interconnected so the connections grow exponentially. To overcome this problem a RBM is model where hidden nodes not linked to one another.

DBN-Deep Belief Networks-Deep Belief Networks is the heap of multi-RBMs on top of each other. In addition, they are connected like this the first RBM outputs becomes the input to the second one and so on it continues. Application areas of DBNs are image recognition, motion-capture and video-recognition and used in medical sector for cancer detection.

DBM-Deep Boltzmann Machines-DBMs structure is almost similar to DBN with one change that is undirected connection between layers. It is fully undirected model. DBMs is suitable for to extracting features that are more complex.

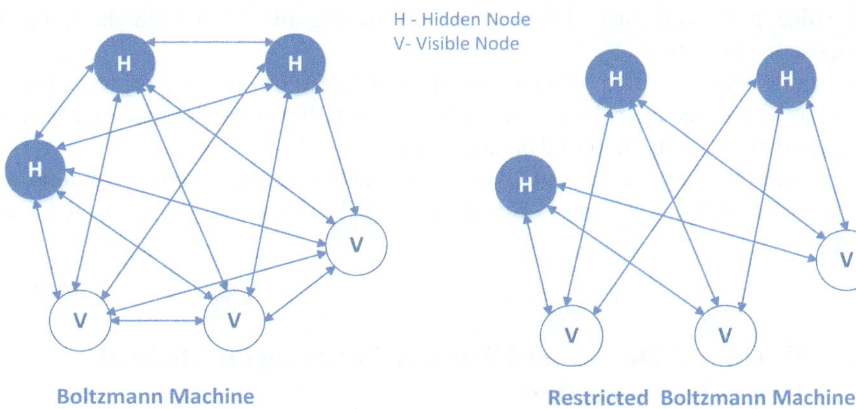

Fig. 11.8 Boltzmann machine and restricted boltzmann machine

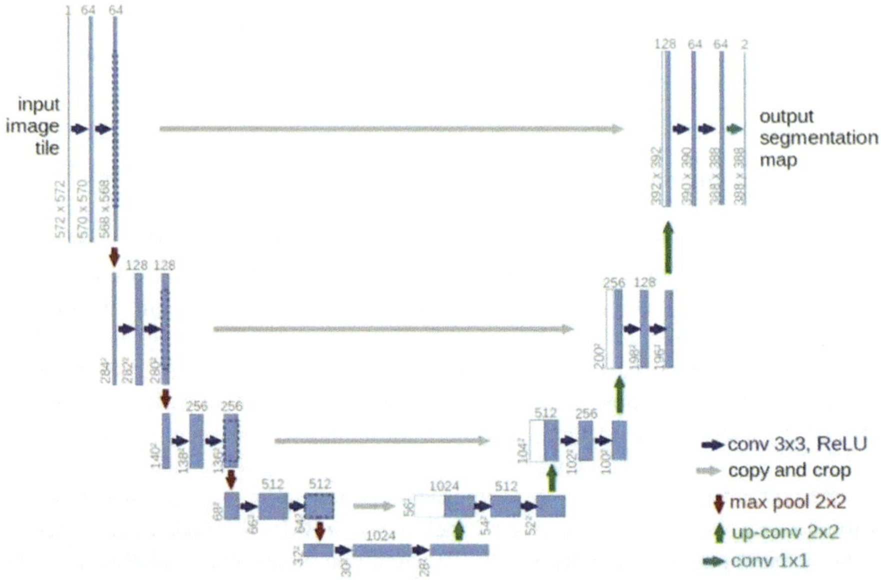

Fig. 11.9 UNet Architecture

11.4.1.6 UNet

The concept of U-net introduced by Olaf Ronneberger et al. [2] comes from traditional CNN, which further rely on the concept of FCN architecture. It is best suited for quick and clear-cut MIS. It uses the theory of inverse convolution and is shown in Fig. 11.9.

The symmetric architecture of UNet has two main parts. First: contracting path (encoder)—comprises with a set of convolutional and max pool process. Second: Expanding path (decoder)—for accurate positioning of object with the help of transposed convolutions.

It can take variable size of image as an input due to the absence of dense layer. UNet model provide pixel-level image classification. It can predict more accurately with less data sets with help of data augmentation.

Different improved variants of UNet are available for getting better spatial information like—3DUnet, UNet++, Residual UNet, Dense UNet, Adversarial U-net etc.

11.5 Review of DL Based Chapters Focusing on Medical Image Segmentation

Table given below shows some work done by researcher for the segmentation of medical images using DL techniques.

Author(s)	Descriptions	Remarks
Oulefki et al. [14]	Target to develop a tool, which automate the process of Lung CT-image segmentation that is very crucial to detect COVID-19 symptoms in very rapid manner. Data set used is COVID-CT-Dataset and EL-BAYANE	Result for proposed tool—accuracy = 0.98, sensitivity = 0.73, precision = 0.71, Dice = 0.71, and specificity = 0.99 which prove it better than its competitors do
Wang et al. [15]	In this chapter, with the help of DL architecture like FCN, U-Net, V-Net etc. they try to build and deploy an AI system. It has features to automate the CT scan image processing process and reduce the human workload in the detection of COVID-19 suspected	This system is able to draw a line between these two types of pneumonia i.e. Common pneumonia & COVID pneumonia. The performance can be improve with adding more training data
Khan et al. [9]	CoroNet model (CNN based) was proposed which helps in spotting corona victim by utilizing chest X-ray. Dataset required- ImageNet and publically available dataset (X-rays of pneumonia)	Results show that still this model need more data to enhance the final accuracy, which is 89.6%. Still it open the door for new researcher
Saood et al. [16]	Purpose of this chapter to compare two architecture, which differ structurally (SegNet vs. U-Net) for investigating diseased tissue areas with dataset of CT scan images	For classification of diseased or un-diseased tissues, SegNet (with 0.95 mean accuracy)-Give better result with respect to other ones. U-Net (0.91 mean accuracy)-Give better performance in multi-class segmentation
Huang et al. [17]	A new UNet 3+ model and hybrid loss function was introduce for automatic segmentation. Best fit for varying scale organs. Reduced network parameters can enhance the computation performance. Data set: images related to liver and spleen	Results-propose model outperforms with hybrid loss function against all cutting-edge approaches. Shows accuracy (Dice liver–0.9675 and Dice spleen–0.9620)
Jha et al. [18]	Architecture ResUNet++, which is modified version of ResUNet and fully automated, was proposed for pixel-wise polyp segmentation Required in detection of Colorectal Cancer (CRC) through colonoscopy examinations Datasets-Kvasir-SEG and CVC-612	The model ResUNet++ results better segmentation for the colorectal polyps with compared toother methods and produces satisfactory results

(continued)

(continued)

Author(s)	Descriptions	Remarks
Wang et al. [19]	AFD-UNet (Adaptive Fully Dense) is neural architecture was proposed. It is a modified UNetmodel by joining the corresponding level in it so that numerous characteristic can be fetched from all layers adaptively Data sets—ADE20K & LiTS (liver cancer) CT-Scan images	The results shows that propose model performs finer than cutting age segmentation methods for the medical images having high variability
Gu et al. [20]	Here a model-context encoder network (referred to CE-Net) was proposed. By adding a context extractor module for fetching notable knowledge while maintaining positional data for segmenting 2D images, is its main feature	Results-proposed method out performs with compare to UNet and brush up the MSI in various functions like lung segmentation, cell contour segmentation etc.
Xu et al. [21]	The main motive—Automatic segmentation of 3D images (like an ultrasound image of the breast) into main working tissues using CNN architecture	This automatic segmentation approach results show good consistency with respect to manual approaches. Able to differentiate between working tissues at pixel level

11.6 Conclusion

The motive behind this study is to highlight the DL models having a great impact on the segmentation of medical images. This chapter initially gives a general idea about medical imaging and segmentation of images and then the review of some chapters introduced the most common deep learning network structures enforced on bio-medical image segmentation by highlighting their advantages. An overview of some important research chapters published in the last decade related to medical image segmentation is presented. After reviewing these chapters, it is noticeable that DL has penetrated every facet of biomedical image analysis, especially for segmentation. With the rapid development in AI technology, DL becomes very popular in researchers choice for medical image segmentation.

Currently, in this COVID-19 pandemic time, detection of the covid-19 positive patient quickly is a challenging task. The researcher has full faith in DL techniques and they are working on developing new architecture based on existing DL models.

However, it looks like CNN and the CNN-based models performers so well in most medical image segmentation tasks, becomes the first choice of many researchers. One noticeable conclusion we can make that no one is exact architecture in getting a good solution.

Each model has its own merits and demerits, so there exist a variety of models to solve different problems. If someone able to analyze the fault of each model, then they can also build new models to improve their models, which fuels the process of innovation.

Researchers can take advantage of this article by selecting the proper DL algorithm as per their problem domain and be ready for coming challenges and related solutions. Finally, chapter comes to focus on this key point that definitely, medical image segmentation has a close relationship with DL.

References

1. T.M. Navamani, Efficient deep learning approaches for health informatics, in *Deep Learning and Parallel Computing Environment for Bioengineering Systems* (2019), pp. 123–137
2. C. Bhatt, I. Kumar, V. Vijayakumar, K.U. Singh, A. Kumar, The state of the art of deep learning models in medical science and their challenges. Multimedia Syst. 1–15 (2020)
3. A.K. Sahoo, K.K. Ravulakollu, Vision based Indian sign language character recognition. J. Theor. Appl. Inf. Technol. **67**(3) (2014)
4. A.K. Sahoo, P.K. Sarangi, R. Gupta, Online platforms for teaching-learning in view of COVID-2019. J. Technol. Manag. Grow. Econ. **11**(2), 49–53 (2020)
5. I. Kumar, C. Bhatt, K.U. Singh, Entropy based automatic unsupervised brain intracranial hemorrhage segmentation using CT images. J. King Saud University-Comput. Inf. Sci. (2020)
6. J. Long, E. Shelhamer, T. Darrell, Fully convolutional networks for semantic segmentation, in *IEEE Conference on Computer Vision and Pattern Recognition* (2015), pp. 3431–3440
7. I. Kumar, C. Bhatt, V. Vimal, S. Qamar, Automated white corpuscles nucleus segmentation using deep neural network from microscopic blood smear. J. Intell. Fuzzy Syst. (Preprint) 1–14 (2021)
8. Y. Bengio, I. Goodfellow, A. Courville, *Deep Learning*, vol. 1 (MIT Press, Massachusetts, USA, 2017)
9. A.I. Khan, J.L. Shah, M.M. Bhat, CoroNet: a deep neural network for detection and diagnosis of COVID-19 from Chest X-ray images. Comput. Methods Progr. Biomed. **98** (2020)
10. R. Yamashita, M. Nishio, R.K.G. Do, K. Togashi, Convolutional neural networks: an overview and application in radiology. Insights Imaging **9**(4), 611–629 (2018)
11. J. Zhai, S. Zhang, J. Chen, Q. He, Autoencoder and its various variants, in *IEEE International Conference on Systems, Man, and Cybernetics* (2018)
12. B.B. Sahoo, R. Jha, A. Singh, D. Kumar, Long short-term memory (LSTM) recurrent neural network for low-flow hydrological time series forecasting. Acta Geophys. **67**(5), 1471–1481 (2019)
13. Z. Feixiang, L. Mingzhe, W. Kun, Z. Hong, Color image encryption via Hénon-zigzag map and chaotic restricted Boltzmann machine over Blockchain. Opt. Laser Technol. **135** (2021)
14. A. Oulefki, S. Agaian, T. Trongtirakul, A. KassahLaouar, Automatic COVID-19 lung infected region segmentation and measurement using CT-scans images. Pattern Recognit. **114** (2021)
15. B. Wang, S. Jin, Q. Yan, H. Xu, C. Luo, L. Wei, W. Zhao, et al. AI-assisted CT imaging analysis for COVID-19 screening: building and deploying a medical AI system. Appl. Soft Comput. **98**(1) (2021)
16. A. Saood, I. Hatem, COVID-19 lung CT image segmentation using deep learning methods: U-Net versus SegNet. BMC Med. Imaging **21**(1) (2021)

17. H. Huang, L. Lin, R. Tong, H. Hu, Q. Zhang, Y. Iwamoto, X. Han, Y.-W. Chen, J. Wu, Unet 3+: a full-scale connected UNet for medical image segmentation, in *IEEE International Conference on Acoustics, Speech and Signal Processing* (2020), pp. 1055–1059
18. D. Jha, P.H. Smedsrud, M.A. Riegler, D. Johansen, T.D. Lange, P. Halvorsen, H.D. Johansen, ResUNet++: an advanced architecture for medical image segmentation, in *IEEE International Symposium on Multimedia* (2019)
19. E.K. Wang, C.-M. Chen, M.M. Hassan, A. Almogren, A deep learning based medical image segmentation technique in internet-of-medical-things domain. Futur. Gener. Comput. Syst. **108**, 135–144 (2020)
20. Z. Gu, J. Cheng, H. Fu, K. Zhou, H. Hao, Y. Zhao, J. Liu, CE-Net: context encoder network for 2D medical image segmentation. IEEE Trans. Med. Imaging **38**(10), 2281–2292 (2019)
21. Y. Xu, Y. Wang, J. Yuan, Q. Cheng, X. Wang, P.L. Carson, Medical breast ultrasound image segmentation by machine learning. Ultrasonics **91**, 1–9 (2019)

Chapter 12
Deep Learning-based Cyber Security Solutions for Smart-City: Application and Review

Tushar Bhardwaj, Himanshu Upadhyay, and Leonel Lagos

Abstract This book chapter provides the readers hands-on experience about the deep learning algorithms and their use case for cyber security solutions in the smart city ecosystem. In this book chapter, the authors first describe the background about the smart city ecosystem, cyber security threats and deep learning algorithms. Subsequently, the authors describe the overview of popular architectures of deep learning algorithms with their applications for addressing various cyber security threats. In this study, the author talks about convolutional neural networks (CNNs), fully connected convolutional networks (FCNs), Recurrent Neural Network (RNN), Deep Belief Network (DBN), Boltzmann Machine, and Auto-encoders.

Keywords Smart City · Deep Learning Algorithms · Cyber Security · Literature Review

12.1 Introduction

"A smart city is a framework, predominantly composed of Information and Communication Technologies (ICT), to develop, deploy, and promote sustainable development practices to address growing urbanization challenges" [1]. In the urban areas of city, there are various types of devices and electronic methods that are connected over the internet/network for the sensory data collection [2–4]. The data includes various end-points of the society such as people, devices, assets, media etc. which is further stored and processed to monitor and analyze important aspects

T. Bhardwaj (✉)
Postdoctoral Associate, Applied Research Center, Florida International University, Miami, FL, USA
e-mail: tushar.bhardwaj@fiu.edu

H. Upadhyay
Principal Scientist, Applied Research Center, Florida International University, Miami, FL, USA

L. Lagos
Director of Research, Applied Research Center, Florida International University, Miami, FL, USA

S. L. Fernandes and T. K. Sharma (eds.), *Artificial Intelligence in Industrial Applications*, Learning and Analytics in Intelligent Systems 25,
https://doi.org/10.1007/978-3-030-85383-9_12

of daily life such as "transportation systems, power plants, utilities, water supply networks, waste, crime detection, information systems, schools, libraries, hospitals, and other community services" [5]. Subsequently, the information is gathered from that collected data which is leveraged to manage the resources, assets and the services efficiently.

The data collected in smart cities is basically is digital form and is quite prone to cyber-attacks. Cyber-attack is basically at attempt to manipulate, expose, destroy, disable, and gain unauthorized access to the computer and its network [6, 7]. Take the case study of transportation data, if that data is being exposed to the attackers the whole city might come to a halt and collapse. Similarly, healthcare data is too critical to be exposed as it may incur severe consequences to the lives of the patients. Researchers are found the cyber-attacks in smart cities are ubiquitous in nature and can happen anytime and anywhere as long as the data is being digitized [8]. In the last decade, artificial intelligence (AI) and machine learning (ML) have encompassed all the important domains of the society. In the lay-man terms, machine learning is "the ability (for computers) to learn without being explicitly programmed".

Machine learning has emerged "boon to the cyber security", which acts as an efficient tool to better analyze and respond to the cyber security threats. Machine learning possesses the ability to go through millions of files and trace potential "hazardous" ones, hence ML is leveraged exponentially to uncover threats and automatically kill them before they can cause havoc.

Deep learning (DL), the subset of ML, which has shown prominent improvements over ML based solutions, as shown in Fig. 12.1. Authors in [9], have proposed a deep learning based intrusion detection system, which is capable of detecting "encryption attacks and zero-day attacks". Authors in [10], proposed an intrusion detection system in the vehicle networking system. Its job is to monitor and detect the real-time network data and detect the cyber-attacks.

This book chapter aim to deliver in-depth details and analysis about the applications of deep learning algorithms for cyber security solutions for smart city. In this study, the authors first describe the background study about the smart city, cyber security threats and deep learning algorithms. Subsequently, the authors describes the overview of popular architectures of deep learning algorithms and used for various cyber security threats. In terms of deep learning algorithms, this book chapter talks about convolutional neural networks (CNNs), fully connected convolutional networks (FCNs), Recurrent Neural Network (RNN), Deep Belief Network (DBN), Boltzmann Machine, and Auto-encoders.

12.2 Background Overview

In this section, the author have highlighted the overview of the three major technologies: smart cities, cyber security and deep learning. This section is intended to relate all these aforementioned technologies and showcases there integration in solving the real world problems.

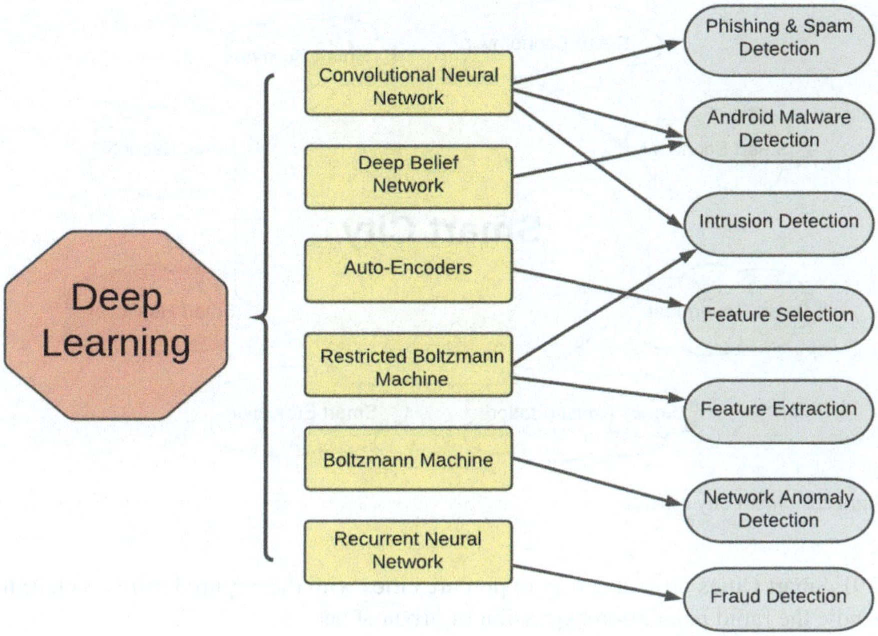

Fig. 12.1 Applications of deep learning algorithms for cyber security in smart city

12.2.1 Smart City

As described by its name, Smart Cities are cities in which technology is used to collect and analyze data from electronic devices and sensors strategically deployed in a city [11], as shown in Fig. 12.2. The goal of gathering the data is to help citizens, private and governmental organizations make decisions based on the data to improve quality of life for everyone [12, 13].

Smart cities can also be described as a framework that integrates the concept of Information and Communication Technology (ICT), in which multiple intelligent devices and machines are able to transmit information to the cloud through wired or wireless connections by using Cloud-based IoT applications [14–17]. Thanks to the IoT devices and data collected, major fields that directly impact the communities can be improved, such as energy distribution, trash collection, police distribution, traffic, etc. These improvements will both greatly decrease costs, while at the same time increasing the quality of life of the citizens [18].

One of the greatest driving forces for Smart Cities is the rate at which urbanization and populations are increasing. As of today, more than 50% of the people throughout the world lives in a city, and it is expected to reach more than 60% in the next 30 years.

With such a large increase in population, there needs to be a way to mitigate the environmental, economic and social needs of the population and cities' resources

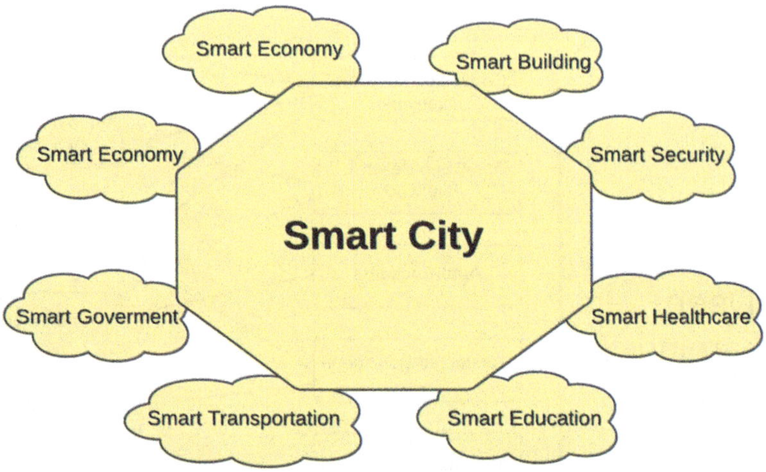

Fig. 12.2 Smart city aspects

[19]. Smart Cities provide a way to prepare cities with the required infrastructure to handle the rapid population expansion of urban areas.

12.2.2 Security of Smart Cities

As technology is introduced to build smart cities with both small- and large-scale applications and devices connected to the cloud, the risk of a cyber-attack increases [20]. When discussing cyber security, there are four objectives that must be integrated by all the parties involved in the deployment and implementation of smart cities solutions: Confidentiality, Integrity, Availability (CIA Model) and finally Accountability. Due to the large amount of data collected through the different sensors and devices, there is a high possibility that some of it will contain information about the consumers and private information. Having a tight grip to prevent unauthorized access to sensitive or confidential information stored in a smart city framework is a must to gain public's trust.

Integrity, on the other hand, refers to the ability of the framework to accurate report the data being collected. The core of Smart Cities is to be able to analyze and take actions on data collected by different devices and sensors. Having the integrity of the data compromised will directly impact the decision and the ability for the Smart cities to properly operate, and steps must be performed to ensure that the data is accurate and unmodified.

Further, Smart cities need to be able to access the data collected in real-time. Ensuring that data is available and accessible by the corresponding agents is crucial for the proper function of Smart Cities. As a result, steps to ensure smooth flow of data between devices, entities and users must be a priority in Smart cities [21].

Finally, Accountability refers to the objective to hold responsible users of a system accountable for their actions. Each action performed on the framework must be logged and directly associated to the user who performed it.

12.2.3 Cyber Security

Cyber security comprises of the practices, technologies, tools and processes that are supposed to protect and prevent the computer networks, data, programs, and devices from the cyber-attacks, such as unauthorized access or intruder attacks [22]. It can also be seen as the information technology security. The importance of cyber security lies in the fact that the huge amount of "unprecedented" data is being generated and used on the computers and other devices [23]. Most of the organizations such as "government, military, corporate, financial, and medical" are intended to generate, collect, analyze, process and finally store their highly confidential data over time. Most of the aforementioned data is quite sensitive in nature, which comprises of the financial data, personal information of the users, medical records [24]. The exposure of this type of data by either being published online or stored in some remote cloud locations leads to the "unauthorized access" and cloud have severe consequences [25, 26, 27]. These organizations are bound to transmit this sensitive data over the computer networks and remote devices for their businesses [28]. Cyber security is aimed at securing the networks and the connected devices and hence protecting the information as well as the systems which are used to process and store the data [29].

Elements of Cyber Security:

- **Identity management:** It talks about the process of managing and handling the access point of each and every individual in the particular organization.
- **Cloud Security:** Most of the sensitive data is being processed and stored in the cloud. Therefore, securing the data over the cloud is quite a huge challenge [30].
- **Network Security:** It is the process of handling the network and protecting it from the unauthorized access, users, intrusions and attacks.
- **Endpoint Security:** As the Covid-19 pandemic hit, remote access is quite necessary point of the business and also the weak point for the data.
- **Mobile Security:** The use of mobile phones and hand held tablets in this virtual world introduces the security challenges and also of themselves.
- **Data Security:** The sensitive data is flowing in all of the networks and connected devices. Ensuring protection to the organizations and customer data is also considered most important and hence require a separate layer of security.

12.2.4 Deep Learning

Deep Learning is the subset of machine learning which is inspired from the structure and orientation of the neurons in the brain cells called as artificial neural networks (ANNs). Artificial neural networks (ANNs) were basically inspired by the "information processing and distributed communication nodes" in the brain. As its name suggests, artificial neural networks tries to mimic the behavior of the actual neurons of the brain. But, technically there are various differences between brain neurons and ANNs such as, the neural networks are more static in nature, on the other hand biological neurons are more dynamic and analogue in nature [31–33].

Although, deep learning was initially coined and formalized in the 1980s, still there are two major reasons due to which it become useful in the recent times:

(a) *Labeled Dataset:* Deep learning takes huge amount of labelled data for model building and predictions. Take the case study of driverless cars, it requires thousands of images and hours of videos to be processed for model building.
(b) *Computing Power:* As mentioned earlier that deep learning model takes millions of record for the training phase, there is the requirement of high performance GPUs with parallel architectures [34–36]. Now-a-days cloud and cluster computing are being leveraged for the model building to minimize the training time from days to few minutes or less [37, 38].

There are various popular deep learning architectures such as "deep neural networks (DNN), deep belief networks (DBN), recurrent neural networks (RNN) and convolutional neural networks (CNN)" [39, 40]. All the aforementioned architectures are applied to various fields in the daily lives such as "including *computer vision, machine vision, speech recognition, natural language processing, audio recognition,* social network filtering, *machine translation, bioinformatics, drug design, medical image analysis,* material inspection and board game programs" [41, 42].

One of the most important application of deep learning algorithms is to detect and prevent the latest malware in the near real-time. The aim of DL algorithms is to minimize the human intervention and also taking care of the "false alarm rate". Deep learning algorithms are also being leveraged by the researchers in classifying the malwares such as "ransomware, backdoors, or spyware" to ensure the security measures in the proactive manner.

12.3 Applications of Deep Learning Algorithms for Safe and Secure Smart Cities

This section showcases the cyber security applications of various deep learning algorithms such as convolutional neural networks (CNNs), fully connected convolutional networks (FCNs), Recurrent Neural Network (RNN), Deep Belief Network (DBN), Boltzmann Machine, and Auto-encoders.

12.3.1 Convolutional Neural Networks (CNNs)

In most of the computer vision technologies, Convolutional Neural Networks (CNN) plays the major role. The CNN leverages the framework with two operations called as "convolution" and "pooling" that minimizes the image size into its important features, as shown in Fig. 12.3. The extracted features are further used to analyze, understand and finally classify the input image.

The CNN architecture has following building blocks:

- *Convolutional Layer:* This is also known as the "filter", sometimes known as "kernel", which is hover over the input image. The primary job of this filter is to traverse pixel by pixel at a particular time frame (example, 4 × 4 filter). The convolution is basically an arithmetic operation which is nothing but the dot product of the input pixel value with the weights of the predefined filter. Finally, the resultant value is the summation into one value that represents the final pixel value.
- *Activation Layer:* The filter operation on the input image reduces its size and hence computes a matrix which is quite smaller than the input image. After that, this matrix is passed through the activation function which allows the non-linearity in the network and enables it to train itself through "backpropagation". Most of the time, the activation function is ReLU.
- *Pooling Layer:* In pooling the image is further downsize/down sampled and hence reduces the matrix size. It basically enables the network to train faster, and also allows the focus on the important parameters in the each feature of the input image.
- *Fully Connected Layer:* It is the traditional convolutional neural network with the input of a one-dimensional vector also representing the output of the previous layer.

Let's discuss some of the common applications of CNN:

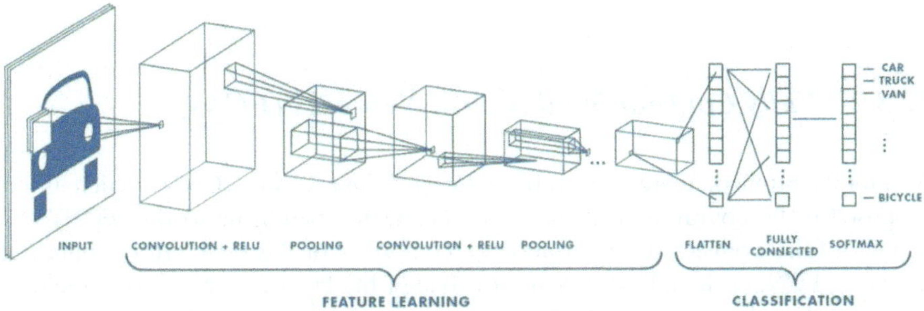

Fig. 12.3 Detailed architecture of Convolutional Neural Networks (CNNs) [43]

- *Healthcare:* The CNN in healthcare is leveraged to detect the image-based diagnosis and detection of diseases like cancer, diabetes and pneumonia. It has been observed that in most of the cases the image based analysis and detection of medical images could be quite accurate as compared to the humans (doctors, physicians and technicians).
- *Agriculture:* In farming, the images of the crops are taken with the help of highly sophisticated "hyperspectral" and "multispectral" sensors. These images are further analysed with the help of CNN to measure its health and finally the viability of the "seed" to be sown.
- *Self-driving cars:* Now-a-days, self-driving cars are most popular the future of the transport industry. CNN is used of object detection and classification on the basis of real-time video streaming generated from the car's cameras. These cars are able to detect people, obstacles and also other vehicles around them.

Authors in [10], proposed a CNN model, which comprises of an input layer, a couple of convolutional layers, a flatten layer, mean pooling layer and also a dense layer. The sole purpose this model is design an intrusion detection system in the vehicle networking system. Its job is to monitor and detect the real-time network data and detect the cyber-attacks. The propose approach was tested on the PC with the accuracy of around 99%.

The study presented in [44], also develops a CNN based intrusion detection mechanism. The proposed CNN model comprises of couple of convolutional layers and pooling layers with the combination of normalization layer for the improvement of the network speed. The model was trained via the pre-processed "historical dataset" to capture the best CNN intrusion detection model. The experimental result shows that the proposed model delivers the average precision of around 0.95 with "Adam optimizer" and number of epochs were 200.

Authors in [45], proposed the CNN based email phishing detection mechanism. For the training purpose, the model used around 6428 emails which includes 4150 "ham emails" and 2278 "phishing emails". The proposed model was tested on the dataset from the "PhishingCorpus" [46], and delivers the accuracy of 99% and the precision of 98%.

12.3.2 Fully Convolutional Neural Networks (FCNs)

As aforementioned, there are "fully connected layers" in a CNN, which is not supposed to be confused with the "fully connected neural networks (FCN)". To define, a fully connected neural network comprises of the set of fully connected layers. In FCN, each and every neuron is one of the layer is connected to the each neuron of the other layer. The primary advantage of the FCNs is that it is "structure agnostic". In other words, there is no need for any particular "assumptions" about the input image. This characteristics of being "structure agnostic", makes the FCNs applications to a very broad spectrum. Also, these types of networks are strong and

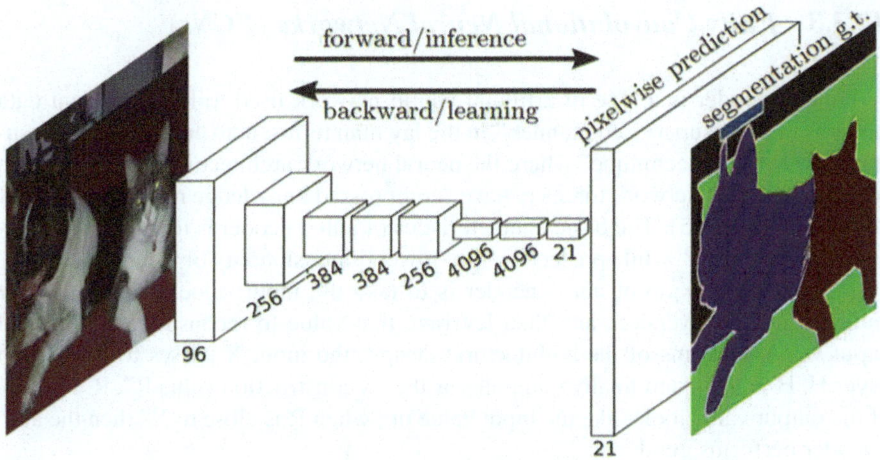

Fig. 12.4 Architecture of FCNs for semantic segmentation [51]

perform very well when fine tuned to a particular problem domain. The change in the architecture of FCNs when compared to the CNNs is that the "fully connected layers" are replaced by the "up sampling and deconvolutional layers" [47]. "The combination of up sampling, deconvolutional and skip connection also led to the development of the deep learning algorithms for various applications" [48–50].

Image segmentation is one of the most common application of FCNs. Which is also leveraged for image-based cancer detection, as shown in Fig. 12.4. The authors in [52] leveraged fully convolutional neural network (FCN) and proposed a distributed but efficient model for the visual "parking lot" detection of the occupancy. The proposed approach is then analyzed and compared with the "state-of-the-art" models by leveraging the two visual datasets, which are "PKLot", and "CNRPark-EXT". In another study [53], the authors presented the surveillance system based on the autonomic prioritization mechanism for the cameras. This study proposed the "selection algorithm", which is inspired from the dynamic channel mechanism of the cameras based on "cognitive radio networks". The experimental results showcase that the proposed model delivers high accuracy for the fire detection scheme which is then compared to the "state-of-the-art" models. Finally, the proposed model also validates the application of the proposed model for the effectiveness of the "fire disaster management". In [54], the authors proposed a unique "PROgressive Vehicle re-ID", known as "PROVID" which is based on deep learning approach. The proposed approach handles the vehicle Re-Id with reference to the two distinct progressive search techniques: "coarse-to-fine search in the feature space, and near-to-distant search in the real world surveillance environment".

12.3.3 Fully Convolutional Neural Networks (FCNs)

"An auto encoder is a type of artificial neural network used to learn efficient data codings in an unsupervised manner." In the lay man terms, auto encoders are "unsupervised learning technique" where the neural network architecture is made in such a fashion that the network forces to have "compressed knowledge representation" of the input image [55]. The basic characteristics of auto encoder is to mimic the "feature representation" while preserving the "lower dimensionality of the input dataset" [56]. The primary job of auto encoder is to take the input value, reduce it to the minimal possible version, and then leverage that value to reconstruct the original input value. In terms of the architectural design, the input X passes to the hidden layer H, H = f(X), and finally computes as the "reconstruction value R", R = g (H). If the output value looks like the input value or "when R is close to X" then the auto encoder performs good.

The architectural design of auto encoder is shown in Fig. 12.5, which comprises of three layers:

- Encoder: This layer compresses the input image into the "latent space representation".
- Code: This layer preserves the "compressed image" being ready for the final decoder layer.
- Decoder: This layer "encoded image back to the original dimension"

Let's discuss some of the real-time applications of auto-encoders:

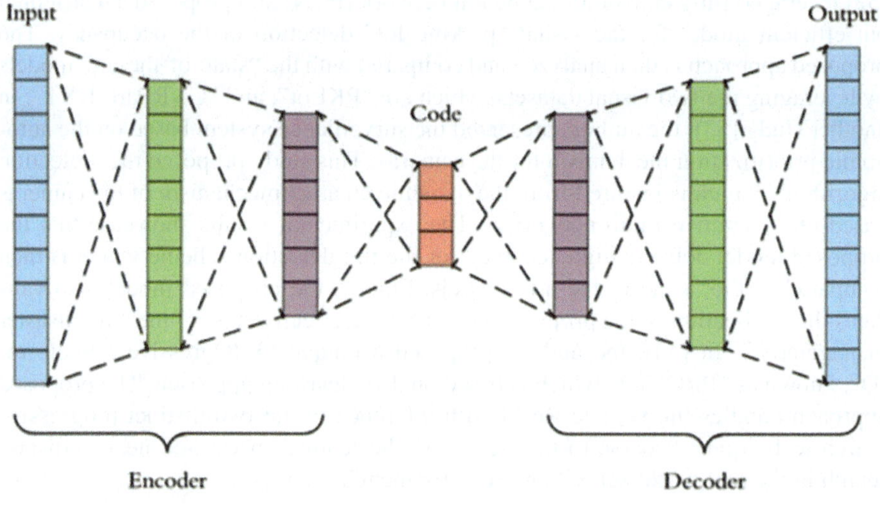

Fig. 12.5 Architecture of Autoencoder [57]

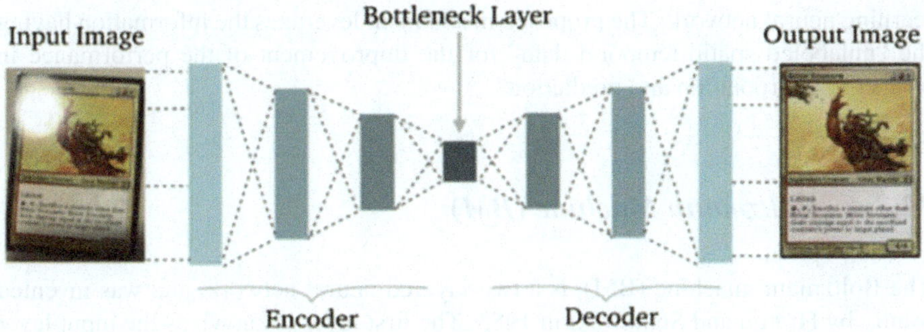

Fig. 12.6 Feature extraction/variance [57]

(a) "Feature Extraction/Variance": It enables in mining the important features from the input image and computes the output image by reducing the noise of non-important information, as shown in Fig. 12.6.
(b) Noise Removal: Auto encoders are specifically designed to take the raw input image which is basically the "corrupted version" on the input image. It is trained to "reconstruct the original image from its noisy counterpart (image)", as shown in Fig. 12.7.

Authors in [59], presented a unique mechanism, which is based on the stacked auto encoder "Levenberg–Marquardt model". The proposed approach is the kind of deep learning architecture of the neural network which is designed to improve the accuracy of the forecasting system. In this mechanism, the "Taguchi method" was leveraged to design and develop an optimized structure. This structure leverages and learns the traffic flow on the basis of "layer-by-layer" granulation of the features "greedy layer wise unsupervised learning algorithm". In [60], the authors proposed a generalized but effective mechanism known as the "Deep Air Learning (DAL)" which solves the three problems. The core of this proposed DAL consists of the "embedding feature selection" and semi-supervised learning for the various layers in the deep

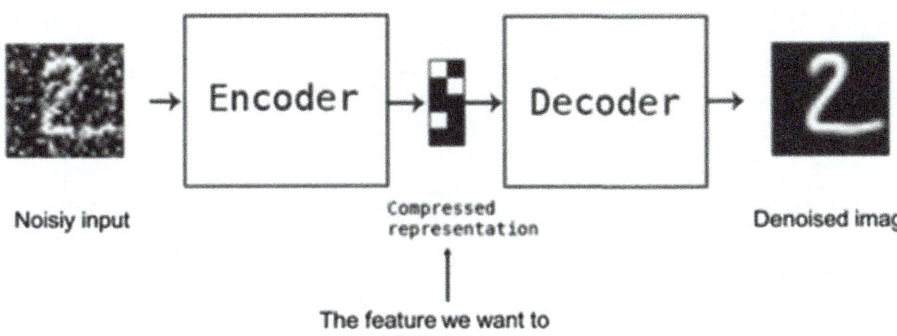

Fig. 12.7 Denoising image [58]

learning neural network. The proposed mechanism leverages the information having the "unlabeled spatio-temporal data" for the improvement of the performance in terms of "interpolation and prediction".

12.3.4 Boltzmann Machine (BM)

The Boltzmann machine (BM), is a two-layered neural network, and was invented jointly by Hinton and Sejnowski in 1983. The first layer is known as the input layer or the visible layer, and the second layer is known as the hidden layer, as shown in Fig. 12.8. There are multiple neurons in each layer. The BM is designed in such a way that there are connections between the neurons of the same and different layers, which is called as "probabilistic neural network". Subsequently, in 2002, Hinton have proposed the "Restricted Boltzmann Machine (RBM)", which is the improvement over the existing BM. In RBM, there are restrictions means the removal of the interconnection between the neurons in the same layer. Therefore, the neurons become independent and have two-way connection between the input layer and the hidden layer.

Authors in [61], leveraged the Deep Boltzmann Machine (DBM) approach for detecting the spoofing attacks in the finger print systems. This DBM model has been proven to have the high accuracy rate of around 86% during the LivDet2013 "spoofing detection competition". The outcome of the research was that DBM is capable of handling complex patterns and promised good results in attack detection.

In other study [62], the authors proposed a unique intrusion detection mechanism which is based on the concepts of feature extraction and dimensionality reduction based on RBM. This study also leveraged the K-means clustering and the KDD-99 datasets for the experimental purposes. The results of this study showcases that accuracy of about 92% for the intrusion detection system.

Authors in [63], presented a modification to the existing RBM by introducing two-layer RBM for the "anomaly network intrusion detection system (A-NIDS).

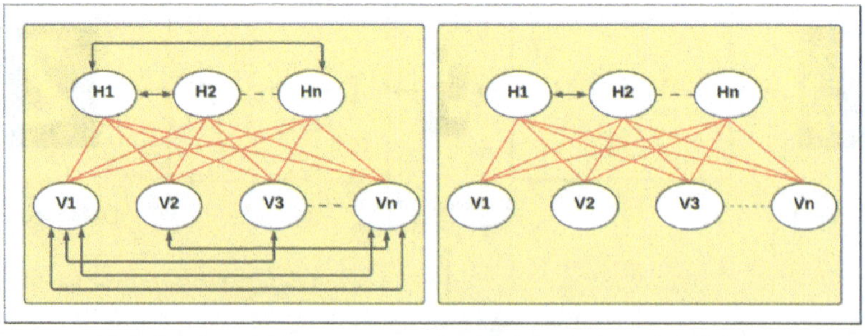

(a) Boltzmann Machine (b) Restricted Boltzmann Machine

Fig. 12.8 Boltzmann machine architecture

This proposed approach can accurately detect the abnormal behavior in the network for the recent "Center of Excellence for Information Security (ISCX) dataset [64]". In addition, this approach also highlights that DRBM could be perform better in terms of "discrimination accuracy" for the real-time cyber security threats.

12.3.5 Deep Belief Networks (DBN)

"In machine learning, a **deep belief network** (**DBN**) is a generative graphical model, or alternatively a class of deep neural network, composed of multiple layers of latent variables ("hidden units"), with connections between the layers but not between units within each layer" [65].

In the year 2006, authors in [66], proposed the "probabilistic generation" model for the DBN on the basis of RBM. The proposed model can be interpreted as deep neural network which is formed by multiple RBM with multiple neurons at each layer, as shown in Fig. 12.9. The DBN model is trained with the combination of un-supervised "pre-training" and supervised "tuning phase".

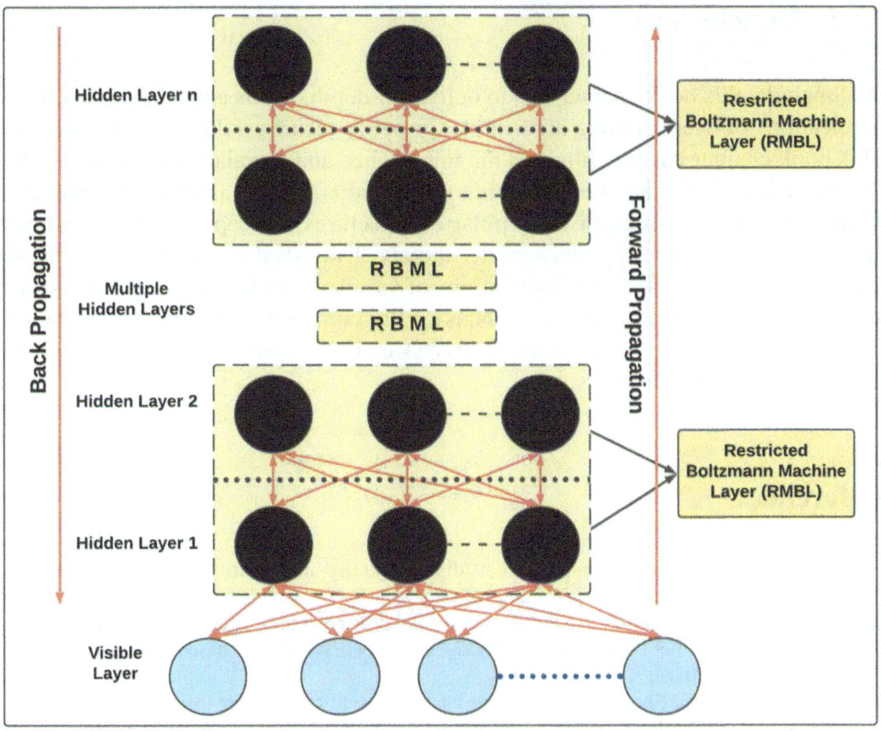

Fig. 12.9 Architecture of deep belief network

This paper [67], leveraged the combination of support vector machine (SVM) and DBM for the real-time detection of the cyber-attacks. To simulate the behavior of real-time detection, generic patterns about the parallel computing technology is used. Subsequently, the DBN is leveraged to transform the original data into the low dimensional dataset. Also, the multi-class SVM are leveraged for the "one-to-all" classification. At last, the "low-dimensional" dataset generated by DBM is leveraged for the intrusion detection by general voting. A study [68], proposed and developed a malicious URL detection mechanism that classifies on the basis of DBN and also the deep neural network (DNN). In this model, the DBM handles the "feature selection", whereas the DNN is leveraged for "binary classification". This proposed approach performs better in training the generation models for the detection of the malicious URL with a good detection rate. The authors in [69], proposed an improved DBN, which is the combination of "Kullback–Leibler (KL)" divergence and non-mean "Gaussian distribution" for intrusion detection. The proposed approach was tested on the public dataset "NSL-KDD [70] and UNSW-NB15 [71]" and have proven that this method have performed far better than the already existing methods for classification accuracy and false positives (FP).

12.4 Conclusions

To conclude, this book chapter aim to deliver in-depth details and analysis about the applications of deep learning algorithms for cyber security solutions for smart city. This book chapter have highlighted the importance and correlation of various cyber security technologies and deep learning applications to form an efficient smart city. It also delivers the overview of popular architectures of deep learning algorithms and use-cases for various cyber security threats. Towards the end, this book chapter talks about the cyber security applications of various deep learning algorithms such as convolutional neural networks (CNNs), fully connected convolutional networks (FCNs), Recurrent Neural Network (RNN), Deep Belief Network (DBN), Boltzmann Machine, and Auto-encoders.

References

1. https://www.thalesgroup.com/en/markets/digital-identity-and-security/iot/inspired/smart-cities
2. T. Bhardwaj, S.C. Sharma, An autonomic resource provisioning framework for efficient data collection in cloudlet-enabled wireless body area networks: a fuzzy-based proactive approach. Soft Comput. (2018)
3. T. Bhardwaj, S.C. Sharma, Fuzzy logic-based elasticity controller for autonomic resource provisioning in parallel scientific applications: A cloud computing perspective, Computers and Electrical Engineering, **70**, 1049–1073 (2018). Author, F.: Article title. Journal 2(5), 99–110 (2016).

4. T. Bhardwaj, S.C. Sharma. Cloud-WBAN: an experimental framework for cloud-enabled wireless body area network with efficient virtual resource utilization. Sustain. Comput. Inform. Syst. **20**, 14–33 (2018)
5. D. McLaren, J. Agyeman, *Sharing Cities: A Case for Truly Smart and Sustainable Cities* (MIT Press, ISBN, 2015)
6. Standardization), ISO (International Organization for. "Publicly Available Standards". standards.iso.org.
7. T.K. Sharma, M. Pant, T. Bhardwaj. PSO Ingrained artificial bee colony algorithm for solving continuous optimization problems", in *Proceedings of the IEEE International Conference on Computer Applications and Industrial Electronics (ICCAIE 2011)*, Malaysia, pp. 108–112
8. F. Li, X. Yan, Y. Xie, Z. Sang, & X. Yuan, A review of cyber-attack methods in cyber-physical power system, in *Proceedings of the 2019 IEEE 8th International Conference on Advanced Power System Automation and Protection (APAP)* (2019), pp. 1335–1339. IEEE.
9. R. Vinayakumar, M. Alazab, K. Soman, P. Poornachandran, A. Al-Nemrat, S. Venkatraman, Deep learning approach for intelligent intrusion detection system. IEEE Access **7**, 41525–41550 (2019)
10. R. Peng, W. Li, T. Yang, & K. Huafeng, An internet of vehicles intrusion detection system based on a convolutional neural network, in *Proceedings of the 2019 IEEE International Conference on Parallel & Distributed Processing with Applications, Big Data & Cloud Computing, Sustainable Computing & Communications, Social Computing & Networking (ISPA/ BDCloud/SocialCom/SustainCom)* (2019), pp. 1595–1599. IEEE.
11. H. Kumar, M.K. Singh, M. Gupta, J. Madaan, Moving towards smart cities: solutions that lead to the smart city transformation framework. Technol. Forecast. Soc. Chang. **153**, 119281 (2020). https://doi.org/10.1016/j.techfore.2018.04.024
12. J. Al Dakheel, C. Del Pero, N. Aste, & F. Leonforte, Smart buildings features and key performance indicators: a review. *Sustain. Cities Soc.* 102328. (2020). https://doi.org/10.1016/j.scs.2020.102328
13. M.P. Efthymiopoulos, Cyber-security in smart cities: the case of dubai. J. Innov. Entrep. **5**, 11 (2016)
14. T. Bhardwaj, H. Upadhyay, and S.C. Sharma. An autonomic resource allocation framework for service-based cloud applications : a proactive approach, in *Proceedings of the 4th International Conference on Soft Computing: Theories and Applications (SoCTA-2019). Advances in Intelligent Systems and Computing (AISC)*, Springer. Scopus Indexed. 27th–29th Dec 2019, India
15. T. Bhardwaj, H. Upadhyay, S.C. Sharma, Autonomic resource allocation mechanism for service-based cloud applications, in *Proceedings of the IEEE International Conference on Computing, Communication, and Intelligent Systems (ICCCIS-2019).* 18th–19th Oct 2019, India
16. T. Bhardwaj, H. Upadhyay, and S.C. Sharma, Framework for quality ranking of components in cloud computing: regressive rank, in *Proceedings of the IEEE 10th International Conference on Cloud Computing, Data Science & Engineering (CONFLUENCE-2020).* 29th–31st Jan 2020, India
17. T. Bhardwaj, H. Upadhyay, and S.C. Sharma, Autonomic resource provisioning framework for service-based cloud applications : a queuing- model based approach, in *Proceedings of the IEEE 10th International Conference on Cloud Computing, Data Science & Engineering* (CONFLUENCE-2020), 29th–31st Jan 2020, India
18. A. Kankanhalli, Y. Charalabidis, S. Mellouli, IoT and AI for smart government: a research agenda. (2019). https://doi.org/10.1016/j.giq.2019.02.003
19. P.T. Tran, T. Nguyen, R. Balasubramanian, Personal exposure to airborne particles in transport micro-environments and potential health impacts: a tale of two cities. Sustain. Cities Soc. **63**, 102470 (2020). https://doi.org/10.1016/j.scs.2020.102470. Author, F., Author, S.: Title of a proceedings paper. In: Editor, F., Editor, S. (eds.) CONFERENCE 2016, LNCS, vol.9999, pp.1–13. Springer, Heidelberg

20. D. Gasper, O.A. G´omez, Human security thinking in practice: 'personal security', 'citizen security' and comprehensive mappings. Contemp. Polit. **21**, 100–116 (2015)
21. K. Bhanot, S.K. Peddoju, & T. Bhardwaj, A model to find optimal percentage of training and testing data for efficient ECG analysis using neural network, Int. J. Syst. Assur. Eng. Manag. (2015). https://doi.org/10.1007/s13198-015-0398-7
22. F. Skopik, G. Settanni, R. Fiedler, A problem shared is a problem halved: a survey on the dimensions of collective cyber defense through security information sharing. Comput. Secur. **60**, 154–176 (2016). https://doi.org/10.1016/j.cose.2016.04.003
23. M. Babar, M.U. Tariq, M.A. Jan, Secure and resilient demand side management engine using machine learning for IoT-enabled smart grid. Sustain. Cities Soc. **62**, 102370 (2020). https://doi.org/10.1016/j.scs.2020.102370
24. M.A. Ferrag, L. Maglaras, S. Moschoyiannis, & H. Janicke, Deep learning for cyber security intrusion detection: approaches, datasets, and comparative study. J. Inf. Secur. Appl. **50**, 102419 (2020). https://doi.org/10.1016/j.jisa.2019.102419
25. T. Bhardwaj, M.R. Pandit, T.K. Sharma, "A Safer Cloud", Data Isolation and Security by Tus-Man Protocol, in *Proceedings of the Second International Conference on Soft Computing for Problem Solving (SocProS 2012), December 28–30, 2012. Advances in Intelligent Systems and Computing*, vol. 236 (Springer, New Delhi, 2014)
26. T. Bhardwaj, T.K. Sharma, M.R. Pandit, Social engineering prevention by detecting malicious urls using artificial bee colony algorithm, in *Proceedings of the Third International Conference on Soft Computing for Problem Solving. Advances in Intelligent Systems and Computing*, vol. 258 (Springer, New Delhi, 2014)
27. T. Bhardwaj, M.R. Pandit. Analysis of cloud security problem and proposed igloo solution, in *Proceedings of the Asia Pacific & MEA Students' Conference, March 14–16, 2012*, (Hong Kong, 2012)
28. M. U˘gurlu, & I.A. Do˘gru, A survey on deep learning based intrusion detection system, in *Proceeding of the 2019 4th International Conference on Computer Science and Engineering (UBMK)*, (2019), pp. 223–228. IEEE.
29. A. Aldweesh, A. Derhab, A.Z. Emam, Deep learning approaches for anomaly-based intrusion detection systems: a survey, taxonomy, and open issues Knowl. Based Syst. **189**, 105124 (2020). https://doi.org/10.1016/j.knosys.2019.105124
30. T. Bhardwaj. End-to-end data security for multi-tenant cloud environment. J. Comput. Technol. & Appl. ISSN: 2229–6964, (2014)
31. A.H. Marblestone, G. Wayne, K.P. Kording, "Toward an integration of deep learning and neuroscience". Front. Comput. Neurosci. **10**, 94 (2016). arXiv: 1606.03813. Bibcode:2016arXiv160603813M. https://doi.org/10.3389/fncom.2016.00094. PMC 5021692. PMID 27683554. S2CID 1994856.
32. B.A. Olshausen, "Emergence of simple-cell receptive field properties by learning a sparse code for natural images". Nature. **381**(6583), 607–609. (1996). Bibcode:1996Natur.381..607O. https://doi.org/10.1038/381607a0. PMID 8637596. S2CID 4358477.
33. Y. Bengio, D.H. Lee, J. Bornschein, T. Mesnard, Z. Lin, "Towards biologically plausible deep learning". arXiv:1502.04156 [cs.LG]. (2015)
34. T. Bhardwaj, M. Kumar, S.C. Sharma, Megh: a private cloud provisioning various IaaS and SaaS, in *Soft computing: Theories and Applications. Advances in Intelligent Systems and Computing*, vol. 584 (Springer, Singapore, 2016)
35. T. Bhardwaj, S.C. Sharma, Internet of Things: route search optimization applying ant colony algorithm and theory of computation, in *Proceedings of Fourth International Conference on Soft Computing for Problem Solving. Advances in Intelligent Systems and Computing*, vol. 335 (Springer, New Delhi, 2015)
36. M.R. Pandit, T. Bhardwaj, V. Khatri, Steps towards web ubiquitous computing, in *Proceedings of the Second International Conference on Soft Computing for Problem Solving (SocProS 2012)*, December 28–30, 2012. Advances in Intelligent Systems and Computing, vol. 236 (Springer, New Delhi, 2014)

37. T. Bhardwaj, S.C. Sharma, "An efficient elasticity mechanism for server-based pervasive healthcare applications in cloud environment", in *Proceedings of the 19th IEEE International Conference on High Performance Computing and Communications Workshops (HPCCWS)*, (Bangkok, Thailand, 2017)
38. K. Kadarla, S.C. Sharma, T. Bhardwaj, A. Chaudhary, "A simulation study of response times in cloud environment for IoT-based healthcare workloads", in *Proceedings of the 14th IEEE International Conference on Mobile Ad Hoc and Sensor Systems (MASS)*, (2017), pp. 678–683. https://doi.org/10.1109/MASS.2017.65
39. J. Hu, H. Niu, J. Carrasco, B. Lennox, F. Arvin, Voronoi-based multi-robot autonomous exploration in unknown environments via deep reinforcement learning. IEEE Trans. Veh. Technol. **69**(12), 14413–14423 (2020)
40. D. Ciresan, U. Meier, J. Schmidhuber, "Multi-column deep neural networks for image classification", in *Proceedings of the 2012 IEEE Conference on Computer Vision and Pattern Recognition*, (2012), pp. 3642–3649. arXiv:1202.2745. https://doi.org/10.1109/cvpr.2012.6248110. ISBN . S2CID 2161592.
41. A. Krizhevsky, I. Sutskever, G. Hinton, "ImageNet classification with deep convolutional neural networks" (PDF). NIPS 2012: Neural Inf. Process. Syst. Lake Tahoe, Nevada. (2012)
42. "Google's AlphaGo AI wins three-match series against the world's best Go player". TechCrunch. (2017)
43. S. Saha, A comprehensive guide to convolutional neural networks — the ELI5 way. *Medium* (2018). at https://towardsdatascience.com/a-comprehensive-guide-to-convolutional-neural-networks-the-eli5-way-3bd2b1164a53
44. W.F. Intrusion detection based on convolutional neural network, in *Proceedings of the 2020 International Conference on Computer Engineering and Application (ICCEA)* (2020), pp. 273–277
45. R. Alotaibi, I. Al-Turaiki, and F. Alakeel, Mitigating email phishing attacks using convolutional neural networks, in *Proceedings of the 2020 3rd International Conference on Computer Applications & Information Security (ICCAIS)* (2020), pp. 1–6. IEEE.
46. Nazario. (2015), *Phishing Corpus* [Online].
47. J. Long, E. Shelhamer, T. Darrell, Fully convolutional networks for semantic segmentation, in *Proceedings of the IEEE Conference on Computer Vision and Pattern Recognition*, (2015), pp. 3431–3440
48. H. Noh, S. Hong, B. Han, Learning deconvolution network for semantic seg- mentation, in *Proceedings of the IEEE International Conference on Computer Vi- sion*, (2015), pp. 1520–1528
49. C. Dong, C.C. Loy, K. He, X. Tang, Learning a deep convolutional network for image super-resolution, in *Proceedings of the European Conference on Computer Vision*, (Springer, 2014), pp. 184–199
50. V. Jain, S. Seung, Natural image denoising with convolutional networks. Adv. Neural Inf. Process. Syst. **21**, 769–776 (2009)
51. What is a fully convolution network? *Artificial Intelligence Stack Exchange* at https://ai.stackexchange.com/questions/21810/what-is-a-fully-convolution-network
52. G. Amato, F. Carrara, F. Falchi, C. Gennaro, C. Meghini, and C. Vairo. Deep learning for decentralized parking lot occupancy detection. Expert. Syst. Appl. **72**, (2016)
53. N. Srivastava, G. Hinton, A. Krizhevsky, I. Sutskever, R. Salakhutdinov, Dropout: a simple way to prevent neural networks from overfitting. J. Mach. Learn. Res. **15**(1), 1929–1958 (2014)
54. X. Liu, W. Liu, T. Mei, and H. Ma, A deep learning-based approach to progressive vehicle re-identification for urban surveillance, in *Proceedings of the European Conference on Computer Vision*, (Springer, 2016), pp. 869–884
55. C.M. Bishop, *Neural networks for pattern recognition*. (Oxford University Press, 1995).
56. A. Ng, Sparse autoencoder, CS294A Lect. Notes **72**, 1–19 (2011)
57. Autoencoder Tutorial. *edureka.co* at https://www.edureka.co/blog/autoencoders-tutorial/
58. D. Bridge, Genetic Algorithms at http://www.cs.ucc.ie/~dgb/courses/tai/notes/handout12.pdf

59. H.F. Yang, T.S. Dillon, and Y.P.P. Chen. Optimized structure of the traffic flow forecasting model with a deep learning approach. IEEE Trans. Neural Netw. Learn. Syst. **28**(10), 2371–2381 (2017)
60. Z. Qi, T. Wang, G. Song, W. Hu, X. Li, and Z.M. Zhang. Deep air learning: Interpolation, prediction, and feature analysis of fine-grained air quality. IEEE Trans. Knowl. Data Eng. (2018)
61. G.B. Souza, D.F. Santos, R.G. Pires, A.N. Marana, & J.P. Papa, Deep boltzmann machines for robust fingerprint spoofing attack detection, in *Proceedings of the 2017 international joint conference on neural networks (IJCNN)* (2017), pp. 1863–1870. IEEE.
62. M.Z. Alom, & T.M. Taha, Network intrusion detection for cyber security using unsupervised deep learning approaches, in *Proceeding of the 2017 IEEE National Aerospace and Electronics Conference (NAECON)* (2017b), pp. 63–69. IEEE.
63. T. Aldwairi, D. Perera, M.A. Novotny, An evaluation of the performance of restricted Boltzmann machines as a model for anomaly network intrusion detection. Comput. Netw. **144**, 111–119 (2018). https://doi.org/10.1016/j.comnet.2018.07.025
64. L. Ghiani, D. Yambay, V. Mura, S. Tocco, G.L. Marcialis, F. Roli et al., Livdet 2013 fingerprint liveness detection competition 2013, in *Proceedings of the 2013 international conference on biometrics (ICB)* (2013), pp. 1–6. IEEE.
65. https://en.wikipedia.org/wiki/Deep_belief_network
66. G.E. Hinton, S. Osindero, Y.W. Teh, A fast learning algorithm for deep belief nets. Neural Comput. **18**, 1527–1554 (2006)
67. H. Zhang, Y. Li, Z. Lv, A.K. Sangaiah, T. Huang, A real-time and ubiquitous network attack detection based on deep belief network and support vector machine. IEEE/CAA J. Autom. Sin. **7**, 790–799 (2020)
68. S. Selvaganapathy, M. Nivaashini, H. Natarajan, Deep belief network based detection and categorization of malicious urls. Inf. Secur. J.: Glob. Perspect. **27**, 145–161 (2018)
69. Q. Tian, D. Han, K.C. Li, X. Liu, L. Duan, & A. Castiglione, An intrusion detection approach based on improved deep belief network. Appl. Intell. (2020)
70. NSL-KDD, (2009). Available from: https://www.unb.ca/cic/datasets/nsl.html
71. N. Moustafa, & J. Slay, Unsw-nb15: a comprehensive data set for network intrusion detection systems (unsw-nb15 network data set), in *Proceedings of the 2015 military communications and information systems conference (MilCIS)* (2015), pp. 1–6. IEEE.

The manufacturer's authorised representative in the EU is Springer Nature Customer Service Centre GmbH, Europaplatz 3, 69115 Heidelberg, Germany. If you have any concerns regarding our products, please contact ProductSafety@springernature.com

Printed and bound by CPI Group (UK) Ltd, Croydon, CR0 4YY
28/09/2026
02224089-0003